The Mystery of
Life Energy

"In this provocative and engaging book, Dr. Leskowitz shares uncensored insights into topics rarely discussed in orthodox medicine. He argues that medicine can only benefit from going beyond its focus on reductionism and the tangible material world. This book will help you understand the apparently paradoxical phrase, 'I'll see it when I believe it.'"

PETER M. WAYNE, PH.D., ASSOCIATE PROFESSOR AND
DIRECTOR OF THE OSHER CENTER FOR INTEGRATIVE MEDICINE
AT HARVARD MEDICAL SCHOOL

"Ahead of the curve and just in the nick of time, Dr. Leskowitz's work is exactly what is needed now if we are ever to harness what is actually true about human potential. Witty and insightful in its transformative revealing of a more enlightened world of healing and forward thinking. This is a must-read on every level."

DR. SUE MORTER, AUTHOR OF THE ENERGY CODES

"When a Harvard Medical School physician tells us something that's absolutely vital to our health and well-being, we should pay close attention. Dr. Leskowitz tells this story, which is firmly grounded in both ancient and modern worldviews and presages a coming revolution in medicine and beyond."

DEAN RADIN, PH.D., CHIEF SCIENTIST,
INSTITUTE OF NOETIC SCIENCES, AND AUTHOR OF REAL MAGIC

"This fascinating introduction to life energy and its healing potential brings together all of the key perspectives on the topic. The author's engaging narrative style, his wry sense of humor, and his half-century of practice as a psychiatrist have given him a bird's-eye perspective that encompasses both conventional and integrative medicine."

DAWSON CHURCH, PH.D., AUTHOR OF
THE GENIE IN YOUR GENES

"Dr. Leskowitz proposes a cartography that will be useful to anyone interested in phenomena on the energy spectrum. These include prehistoric stone circles, phantom limbs, sports events, rock concerts, and people exuding charisma. He concludes by explaining 'how it works,' a true tour de force."

STANLEY KRIPPNER, PH.D., AFFILIATED DISTINGUISHED FACULTY
AT THE CALIFORNIA INSTITUTE OF INTEGRAL STUDIES

"Dr. Leskowitz has dropped a big rock in the pond, and the ripples are making waves. By building upon considerable scientific evidence, Dr. Leskowitz's 40 years of work has dramatically pushed out integrative healing practices for chronic pain management, even in the NFL. A new day is happening!"

DAVID MEGGYESY, EX-NFL LINEBACKER, AUTHOR OF *OUT OF THEIR LEAGUE*

"Dr. Leskowitz elegantly discusses a broad spectrum of human energy topics with irreverent humorous stories and brilliant scientific insights that encourage us to embrace a new metaphor for healing based on an energetic paradigm. If you want to know how to make your heart frequencies more coherent and get into the zone of peak performance, this book is for you!"

LARRY BURK, M.D., CEHP, AUTHOR OF *LET MAGIC HAPPEN*

"An excellent book that includes many extraordinary aspects of life energy and its relevance to science and medicine. Ancient mystics understood and modern physicists know that all is vibration. Now you can too! Whether you're new to the subject or have been studying it for a long time, *The Mystery of Life Energy* has something in it for everyone."

JONATHAN GOLDMAN, AUTHOR OF *HEALING SOUNDS*

"A grounded guide to sovereign self-management inside the broader context of the global consciousness elevation. If you're looking to elevate your life energy management, you'll find valuable wisdom inside this book."

DAVID GRUDER, PH.D., FOUNDING PRESIDENT OF
THE ASSOCIATION FOR COMPREHENSIVE ENERGY PSYCHOLOGY

"A seriously good book, inviting you to look at the many faces of life energy, literally. A guide-by-example to interesting aspects of consciousness and how to learn from them."

ROGER NELSON, DIRECTOR OF THE GLOBAL CONSCIOUSNESS PROJECT

The Mystery of
Life Energy

Biofield Healing, Phantom Limbs, Group Energetics, and Gaia Consciousness

Eric Leskowitz, M.D.

Bear & Company
Rochester, Vermont

Bear & Company
One Park Street
Rochester, Vermont 05767
www.BearandCompanyBooks.com

Text stock is SFI certified

Bear & Company is a division of Inner Traditions International

Cataloging-in-Publication Data for this title is available from the Library of Congress

ISBN 978-1-59143-486-3 (print)
ISBN 978-1-59143-487-0 (ebook)

Printed and bound in the United States by Lake Book Manufacturing, LLC
The text stock is SFI certified. The Sustainable Forestry Initiative® program promotes sustainable forest management.

10 9 8 7 6 5 4 3 2 1

Text design by Kenleigh Manseau and layout by Debbie Glogover
This book was typeset in Garamond Premier Pro with Futura Std, Gill Sans MT Pro, and Manofa used as display typefaces

The author made every effort to locate and contact rights holders for all images used to obtain permission and give credit.

To send correspondence to the author of this book, mail a first-class letter to the author c/o Inner Traditions • Bear & Company, One Park Street, Rochester, VT 05767, and we will forward the communication or contact the author directly at **themysteryoflifeenergy.com**.

Scan the QR code and save 25% at InnerTraditions.com. Browse over 2,000 titles on spirituality, the occult, ancient mysteries, new science, holistic health, and natural medicine.

Contents

PART IV
Large-Scale Manifestations of Life Energy

Foreword

David Feinstein, PhD

My wife, Donna Eden, and I first learned of Rick Leskowitz in 1999 when he reviewed our book, *Energy Medicine*, for a prestigious integrative medicine journal. He gave it a glowing appraisal and endeared himself to us forever when he revealed that his primary disappointment with the book was that *he* didn't write it. Now, a quarter century later, he has written his own energy medicine book, and you are in for an exciting adventure.

We wondered, who is this board-certified psychiatrist who has apparently penetrated behind the enemy lines, holding a position at Harvard Medical School and bringing energy medicine techniques into a major Harvard teaching hospital. Of course, by now, conventional medicine is no longer the "enemy" as energy medicine is being more widely embraced within traditional institutions, and it is partially through Dr. Leskowitz' tireless dedication that this shift is occurring. It would be easy to underestimate the pioneering courage required to bring the widely ignored or disparaged methods he was using back then into one of the world's top medical institutions.

The Mystery of Life Energy, as I see it, delivers on three major aspirations:

1. It weaves into a single tapestry a wide range of disparate and fascinating phenomena, all tied together with the bright thread of life energy, and in a manner that will give some refreshing jolts to the reader's perspective.
2. It unflinchingly examines each of them from, in the author's words, "both sides of the aisle": conventional linear science and traditions that are grounded in an appreciation of the "multidimensional nature of reality." Few scientists are so deeply grounded in both sides of this gaping chasm (it's not *just* an aisle) as to allow them to seriously attempt such a feat. Dr. Leskowitz does it masterfully.
3. In the process, you are taken on a journey in which one of the core paradigm clashes of your culture is piece-by-piece disassembled, put under the bright lights of reason and experience, and reconstructed in a way that can move the culture—as well as your intellect—forward, informed by both sides of the chasm.

The author's credential that most impresses me is his application of energy techniques to effectively relieve suffering and promote healing. That's not to take anything away from my fascination with the other topics in the book, but it is where the proverbial rubber meets the road, regardless of how conventional or outside the lines the method might be. The outside-the-lines topics in this book, however, are likely to trigger your interest or at least your skepticism.

What tempered my own skepticism around some of the controversial topics, such as crop circles and precognition, was the conversational style where the author's amazement is woven into the ways he makes sense of his own experiences, of validated reports, and of scientific research that all truly stretch the mind. Through this somewhat autobiographical feature of the book, we see how Dr. Leskowitz's own evolution mirrors the change that society is experiencing as we all go through a paradigm shift.

The culture's resistance to this shift is examined in a series of case examples that describe how imprisonment, internet censorship, and even book burning were used as tools of control. We see how Franz Anton Mesmer was indelibly branded as a charlatan and is still mocked despite continuing evidence of the validity of many of his core ideas. Wilhelm Reich died in prison for distributing a machine that purportedly worked with the body's healing energies. In 1998, the American Medical Association "debunked" the energy medicine practice of *Therapeutic Touch* in its prestigious journal based on surprisingly amateurish data. And, when provided with irrefutable feedback that Wikipedia has consistently ignored efforts to get it to mention that over one hundred clinical trials demonstrate the effectiveness of energy psychology, its founder resorted to childish name-calling instead of lifting the censorship. The inertia of the materialistic worldview and the relationship between science and politics are illuminated throughout the book.

Out of the smoldering ashes of outdated models within established institutions, a phoenix of new and vital breakthroughs in our understanding is rising. *The Mystery of Life Energy* chronicles these developments. I had to rethink my dismissal of the more far-out explanations for crop circles. In another area, I have no intellectual context for reports throughout history of precognition, of knowing what is going to happen before it occurs. So what am I to do with the puzzled reflections of an NFL linebacker who "could sense the movements of the running backs a split second before they happened?" Mind-stretching examples permeate the book. Don't miss the breathtaking implications for global consciousness in the final chapter.

The book's scientifically informed perspective combines with down-to-earth stories that stretch or counter conventional thinking. All this is leavened with the author's unique sense of humor and irony. And it's done in a manner that makes a potentially esoteric topic eminently approachable. So have a go at it, and let your boundaries be stretched a bit. You're in good hands, and I think you'll enjoy the ride.

DAVID FEINSTEIN
ASHLAND, OREGON

DAVID FEINSTEIN, PHD, is a clinical psychologist and nine-time national award-winner for his books on consciousness and healing. He and his wife, Donna Eden, established the world's largest hands-on organization teaching energy medicine, and their book, *The Energies of Love,* achieved bestseller status on the New York Times Relationship List. He has also written more than one hundred articles in professional literature. Feinstein has served on the faculties of The Johns Hopkins University School of Medicine, Antioch College, and the California School of Professional Psychology.

INTRODUCTION
Shifting Perspectives

WHAT'S THE PROBLEM?

"The heart is a pump, the eye is a camera, and the brain is a computer."
These three metaphors are at the core of every American high school
biology class and are used to describe the function of various organs in
the human body. They're certainly helpful as far as they go, but I don't
put these images in the one-picture-is-worth-a-thousand-words category.
In fact, I think they are wrong turns that have sent modern science and
medicine (and even society itself) off in a mistaken and even dangerous
direction. Of course, the metaphor of the human body as a complex
machine is enticing and has led to many valuable medical and surgical
advances. But it is a fundamentally shortsighted metaphor because it
completely overlooks the essential component of being alive—the sea of
invisible and intangible life energy that we literally swim in, the reservoir of vitality that infuses us with profound resilience. It is our most
direct connection to the world at large and to each other, a connection
we have been ignoring at our peril.

In fact, every healing tradition the world over—except for modern
Western medicine—invokes the concept of an invisible healing force or
energy. Many of these foreign terms for energy-based practices have even
found their way into contemporary American culture—the Chinese
exercise of tai chi, the qi that is stimulated by acupuncture needles, the

yoga postures and mantra meditations, and so on. But Western medicine hasn't yet embraced the idea that something invisible, apart from chemicals and cells, gives us our human essence. As this book will demonstrate, this disconnect is responsible in large part for many of society's crises today—medical, interpersonal, ecological, and geopolitical.

Many writers have already described in detail the shortcomings of this mechanical, materialist worldview.

What I hope to add to the conversation is my perspective as a conventionally trained physician who is also a long-term practitioner of mind/body medicine, and a student of energy healing for over thirty-five years (but I had opted for surgery rather than meditation and acupuncture to repair my torn knee ligament years ago). The goal of this book, then, is to describe a completely different point of view (POV) that emerges once we acknowledge that "energy" is real. And life as a whole looks very different when viewed through this lens.

HOW MY PERSPECTIVE CHANGED

In addressing this energy-centered POV, I'll work both sides of the aisle. I'll balance a discussion of scientific advances and research findings with personal stories and experiences, clinical reports, and practice exercises for the reader. I'll comment on cultural trends in science and society to reveal how a new medical worldview opens the door to a cultural shift in perspective. In this way, the hard analytic data favored by the brain's left hemisphere will be complemented by images, emotions, and experiential practices that are at home in our impressionistic right hemispheres. Hopefully, this synthesis will provide a catalyst for change—in the reader and in the culture.

To begin, it's important to put my cards on the table and make a confession: I was brought up in the tradition of scientific materialism—not in the sense of lusting after the latest consumer goods, but as a believer that life's essence is found in material structures, that every human experience can be reduced to matter, and that the only trustworthy sources of information about the world are our five physical senses. I was the son of a scientist, an organic chemist turned immunologist, and I loved the excitement of the real-life not-on-TV microscopes and test tubes in my father's research lab. However, I didn't realize that while I was being entranced by these powerful tools, I was also being indoctrinated into a very limited worldview. In school, I was good at math and science, so I got more than my fair share of approval and ego gratification for pursuing those interests. Studying medicine seemed to be the best way to harness and develop those skills while helping others in the bargain. I never thought too deeply about the big questions: the nature of illness, where consciousness came from (the brain, obviously!) or what love was (I hadn't fallen in it yet), and so I looked forward to going to medical school and getting all the answers.

Fortunately, I made a crucial decision before starting that training. After college, I took a couple of years off to travel in Europe, the Middle East, and South Asia. I saw enough of non-European cultures to realize that our Western approach was just one of many ways of looking at reality, each with their own integrity and validity. So, when I began med school in 1975, I had already gained enough perspective on the subject to suspect that my professors might not have as complete an understanding of health and illness as they professed. For example, I was disappointed to learn that many diseases were considered to be "idiopathic" (a term that was Greek to me—literally—until I found it was just a fancy way of saying that the cause wasn't known), while common disorders like hypertension were "essential" (in other words, they just happened). There were no explanations to satisfy my sense of curiosity, but plenty of details to memorize (we were told early on that at least half of the material we were being taught would be obsolete in ten years, but since the professors didn't know which half was which, we should learn it all!).

My classmates and I joked that Latin or Greek should have been a required course for pre-med students because so much of the mystique of medicine depended on the use of an impenetrable Greco-Latinate jargon. Would you have more confidence in a doctor who told you that he didn't know why your skin was turning purple and you were running out of blood clotting cells, or in one who told you that you had a classic case of idiopathic thrombocytopenic purpura? That mumbo jumbo is a direct translation into Medicalese of the Anglo-Saxon description.*

Bottom line: no one really seemed to know *why* things happened to our patients or why they happened at the particular time that they did. So they filled in the causal blanks by burrowing down into ever more microscopic levels of understanding. Genes were always the fall-back explanation, but that left us feeling like we were just robots playing out the genetic hand we were dealt—not at all satisfying to my wish to understand at a causal level how health and illness were generated, where they came from, and how health could be sustained.

As a compromise, I decided to specialize in psychiatry because it had a different level of focus than straight internal medicine. At least psychiatrists talked about thoughts and emotions without trying to reduce them to chemicals (this was just before psychopharmacology and medication management took off as the field's raison d'être). I was fortunate enough to train with some of the founding fathers of psycho-neuroimmunology (PNI), Greco-Medical for "the study of mind, the nervous system and the immune system." PNI was a definite upgrade from earlier psychoanalytic notions of "psychosomatic," usually taken to mean that it's all in your head (ie, imaginary). PNI was the thin end of the wedge because it showed how stress affected all of our body's biologic functions, including immunity. That's a "D'uh!" insight today, but forty-five years ago it was a radical notion. So in order to prove that these insights were legit, and to gain acceptance in the halls of academic medicine, PNI became focused on reducing these behavioral patterns to hormones and neurotransmitters and nervous system responses, the sort

---

*Here are the Greek stubs, followed by their meaning, and in parentheses a modern English word that comes from the same root: *Idio-*: unknown (idiot); *-pathic*: disease (pathetic); *thrombo-*: blood clot (lump); -cyto-: cell (hut); *-penic*: lack of (penury); *purpura*—bleeding under the skin (purple).

of structures my medical colleagues had faith in. To be fair, they also believed in energy, but it was ATP (adenosine triphosphate), the chemical fuel that made cells go.

It wasn't until 1985, when a nonphysician colleague invited me to a lecture on "energy healing," that things really began to fall into place (or fall apart, depending on your POV). The term "healing" evoked images of "the laying on of hands," which was too off-puttingly Christian even for my nonobservant but culturally Jewish upbringing. I was fortunate enough though to hear an extraordinary speaker, the Rev. Rosalyn Bruyere, an LA-based healer who had earned the respect of her local medical community for the simple reason that she got results. Most puzzling to me, she had no formal medical training, having studied to be an engineer. Instead, she was able to assess the clinical situation using what she called her clairvoyance, her intuitive inner vision. Despite having my intellectual defenses triggered by all these red flags, it took all of ten minutes into the lecture for me to be hooked. She clearly knew her stuff and was talking about the missing ingredient that conventional doctors had been overlooking—the energy that linked thoughts, emotions, and the body. Nothing seemed idiopathic any more.

For the next seven years, I attended her regular healing intensives in Western Massachusetts, where about forty of us regulars learned the vagaries of energy healing, shamanism, spiritual growth, and the multidimensional nature of reality. I also learned, to my dismay, that intellectual prowess was not the be-all and end-all in this line of work. In retrospect, one of the most important lessons I learned from those workshops was humility—for the first time in my life, I was the dumbest kid in class. I spent so much time in my head that I couldn't feel the energy moving through my body that my classmates were busy sensing and modulating. My main challenge was learning how to put aside the steady stream of thoughts, judgments, and analyses that had been so valuable in my medical training and instead tuning into the subtle messages from my body's inherent energy-sensing systems.

I soon realized that it was helpful to bring these energy-based approaches into my regular pain management practice, where I was part of a multi-disciplinary team that taught self-management skills to chronic pain patients. In that way, my patients might benefit from the healing forces I was learning about. But once I started looking at the world through the lens of invisible healing energies, everything shifted. It wasn't just that new clinical options opened up for my patients, but also other seemingly unrelated hobbies and interests began to make sense. As I would come to see, energy was the common denominator (or as they say in med-speak, the final common pathway) for everything from pain management to athletic performance, from team chemistry and fan energy to sacred sites, paranormal phenomena, and global consciousness. Those connections and their energetic underpinnings have been the focus of my work for the last thirty-five years.

THE PATH AHEAD

In part I of our journey, we'll use several case studies to look at the history of Western medicine's longstanding antagonistic relationship to healing energies, starting with the notorious father of animal magnetism, Franz Mesmer, and building toward a final example of social media as a powerful force for energetic suppression. But the times are changing, and we'll also see how the field of energy medicine is gradually gaining acceptance in academia and the medical mainstream. My own bumpy road through the halls of academia will illustrate this transition.

Entering the clinical realm, part II will describe the huge difference energy-based treatments made for my chronic pain patients, allowing them to shift away from reliance on medications toward self-management. This was not a placebo effect: many pain conditions stemmed from imbalances of energy flow that appeared to mimic nervous system disorders. Some conditions, like the bizarre syndrome of phantom limb pain, even provided an opportunity to directly perceive and confirm the existence of energy fields. When I discuss this syndrome in more detail later, readers will learn how to detect their own biofields by using a simple adaptation of a common energy therapy, forcing the standard paradigm of Western medicine to give birth to a broader view. Medications and surgery can certainly do wonderful things (e.g., my knee), but it's time to expand into new dimensions.

In the final two sections, I'll make the leap from individual energy effects to group energy interactions. This larger-scale phenomenon became evident to me because I was a rabid sports fan, having grown up in the Greater Boston area (where sports fanaticism comes with the territory). While following my favorite teams and players, I had noticed that the energetic aspects of athletic performance were obviously heightened during those times when players were in the proverbial Zone of peak performance. By taking a look at the energetics of being in the Zone, whether you're an athlete (elite or amateur), musician, public speaker, or just an average person doing what you do each day, we can learn how to be, in the words of tennis coach and colleague Scott Ford, "in the Zone by choice, not chance."

But even more telling than these individual energy shifts were the intense energies generated by the entire sporting event, by the players as a team, and especially by the fans in the ballparks. This growing awareness coalesced for me in 2004, when our local baseball team, the Boston Red Sox, won the World Series for the first time in eighty-six years. I was struck by how everyone in New England was so emotionally invested in their championship run, whether a fan or not, that the energies of the entire region seemed to be elevated together.

My filmmaker cousin and I eventually told this story as a documentary film. The Red Sox organization allowed us to interview players about how they responded to fan energy and created team chemistry. We also spoke with leading "Hall of Fame" energy researchers. We used state-of-the-art computer software to measure these group energy effects in the lab and in the ballpark, and our film—*The Joy of Sox: Weird Science and the Power of Intention* (2013)—was broadcast on PBS.

A spin-off group of coaches, athletes, and sports psychologists continues this work by bringing ideas of energy and consciousness to mainstream culture, where a national newspaper article about a pro football line-backer's yoga practice will probably have more of an impact on cultural evolution than any five randomized, controlled clinical trials conducted by the NIH (the National Institutes of Health).

Another seemingly unrelated long-term passion became much more understandable when I considered the role of invisible energies in mystical places like the ancient stone circles of England. Under the guidance of my British wife, I have visited England many times and walked among the great standing stones of Avebury and Stonehenge (back in the days when the public could actually touch the stones, before the latter was fenced off). And the sense of mystery and wonder in these spots was palpable. But then, starting in the 1990s, we began to hear about huge, complex geometric designs that appeared overnight in the wheat fields of Southern England, patterns that could only be seen from the skies above. These mysterious "crop circles" soon became a personal obsession, and before long I was convinced that they were not hoaxes (as the mass media wanted us to believe), but intelligently designed patterns that appeared in that particular part of England for good reason.

I learned about geomancy—the study of Earth energies, ley lines, and sacred sites—and I came to realize that the circles were yet another forum in which invisible energies could work their magic. Gaia—the classic Greek goddess also known as Mother Earth—had her own energy system, with exact analogues to acupuncture points and meridians, and these circles only appeared in areas where her energy pathways were optimally aligned. The evidence also suggested that these complex and geometrically precise crop circle designs did not originate on our planet, but that extra-terrestrial intelligences were involved in their creation. Needless to say, this was the deepest of all my rabbit holes!

Finally, it turned out that group energies could also be measured at a distance—around the globe, in fact. The same software that detected changes in fan energy levels inside Fenway Park could also measure consciousness shifts at the global level. A worldwide network of these devices—The Global Consciousness Project—had been taking the pulse of our world "brain" since 1998. Their results showed that we are all interconnected by yet another form of invisible energy, giving scientific credibility to such previously hypothetical concepts as Carl Jung's *collective unconscious* and Teilhard de Chardin's *noosphere*. We were finally developing empirical explanations for paranormal experiences like telepathy and synchronicity.

If all of this material is as valid as the evidence suggests, then we have some powerful new/old tools to help humanity deal with its current lineup of catastrophes. We can repair our damaged ecosystems by revitalizing Earth's energy grid, we can resist a wide range of illnesses (including Covid) by using energy medicine techniques to enhance our natural immune resistance to disease, and we can reverse the energetic costs of social isolation by remembering and re-experiencing the web of invisible energy that connects and nourishes us all. When we acknowledge that we're interconnected energy beings rather than separate machines, we can harness our innate healing gifts and benevolent interdependence to find a non-polarizing way forward that works for everyone.

So that's what lies ahead. At one level, it's the story of how a series of seemingly unrelated personal interests triggered one person's change from neuroscience nerd to energy adept. But more importantly, it's an outline of our current cultural shifts as we seek to birth a new paradigm that has room for both science and mysticism, a paradigm that can guide us to a future where we can all thrive together.

Glossary of Acronyms

This book uses many acronyms both for professional organizations and brevity. At the back of the book you will find a Glossary of those acronyms for your reference.

PART I

A Brief History of Energy

1
Emerging
Views of Energy

West Meets East

East is East and West is West, and never the twain shall meet.

RUDYARD KIPLING,
THE BALLAD OF EAST AND WEST (1889)

THE STORY OF FRANZ MESMER:
THE QUACK VS. THE KING

"Charlatan, fraud. . . or genius?" That's not a multiple-choice exam, but the headline of the publicity poster for a 1994 biopic about Franz Anton Mesmer, the eighteenth-century German doctor whose discovery of animal magnetism turned the medical world upside down. Although his public image has migrated toward the charlatan/fraud end of the reputational spectrum since then, this chapter will make the case that he was in fact a genius of sorts, and that his work set the stage for what I hope will be the eventual acceptance of healing energies by Western medicine and society as a whole. Ironically, the same political forces and intellectual biases that forced Mesmer to leave Paris in disgrace 250 years ago more recently led the prestigious *Journal of the American Medical Association* (JAMA) to try to debunk the notion of energy fields by publishing a study that was written by an eleven-year-old school girl! So those forces are important to study.

The story of his rise and fall illustrates so many key points about how paradigms shift that it provides an excellent template for understanding sociopolitical processes that are still in operation today. Here are five key aspects of the process that will guide us as we take a closer look at Mesmer's life and the challenges that he, and all innovators, face in developing a conceptual model that conflicts with accepted scientific views:

- The innovator's personal path of inspiration and discovery
- The conservative and dogmatic nature of conventional medicine
- The financial and political pressures that hinder the development of new medical treatments
- The role of the Establishment in controlling the dissemination of information
- The way that historic legacies are shaped by The Powers That Be

As with other great innovators, Mesmer's origin story is important. Franz Anton Mesmer was born in 1734 in rural Germany, near Zurich, Switzerland. The son of a forester, he went to a Jesuit university and then to medical school. His medical thesis was entitled "On the Influence of the Planets on the Human Body," a form of medical astrology that assessed the "tides" within the body that were generated by the movements of the planets, in particular the moon and sun (NB: Medical Astrology was not part of the curriculum when I attended medical school, not even as an elective). But in a rather impressive sign of how the pendulum has swung back again, at least one Harvard-affiliated psychiatrist now offers this diagnostic option in her practice (Tsafrir 2024). Mesmer married a wealthy Viennese widow and set up his practice in this epicenter of Europe's Age of Enlightenment at a time when the scientific world was abuzz with the latest discoveries about the mysterious forces of electricity and magnetism.

Mesmer's inspirational idea was to harness these new developments into an approach to medical treatment by having his patients ingest metallic salts and then placing lodestones (naturally occurring magnetic rocks) on their body to augment their inner tides. He soon came to believe that he could enhance the process by transmitting his own personal magnetic force through his palms, which he stroked several inches from the surface of his patient's body, in so-called mesmeric passes. He called this biologic healing force animal magnetism, as a naturalistic complement to electromagnetism (Ellenberger 1970).

He established a flourishing practice in Vienna, but a very public failure in treating a well-known blind pianist led him to leave his wife behind and move to Paris. At this time, many improbable scientific breakthroughs were regularly publicized and popularized, including the invention of "elastic" shoes that would allow one to walk on the Seine River. Another amateur scientist claimed to be able to breathe underground and burrow to his destinations (Steptoe 1986). In this climate, Mesmer's relatively more realistic work with actual physical magnets became so popular that he devised a form of group treatment in order to accommodate his growing clientele of wealthy aristocratic patients. (His business model was also aided by the fact that the physician to King Louis XVI's brother was himself a convert to mesmerism.)

Modeled after the newly invented Leyden jar, the first capacitor to store electricity, Mesmer used large wooden basins (*baquets*) filled with water that he claimed to have previously energized with his personal magnetism; they were literally capacitors for human magnetism. A series of iron rods emerged from the water around the basin's perimeter and could be held by up to twenty patients at once, allowing many people to receive simultaneous magnetic charges (up to four baquets were in operation at a time). This proved to be such a good business model that the other physicians in Paris grew concerned about his growing competitive advantage.

Mesmer was popular enough to come to the attention of such well-known historical figures as Thomas Jefferson and the composer Mozart. Jefferson was the U.S. minister to France in the 1780s, before their Revolution, and he attended a baquet session with Mesmer as portrayed in the 1995 Merchant/Ivory film *Thomas Jefferson in Paris*. Jefferson is shown as a lonely and somewhat repressed widower who is taken aback by the moaning and groaning coming from the aristocratic Parisian women as they fall under Mesmer's spell—aided, no doubt, by Mesmer's Harry Potter-ish magic wand and his famous lilac silk robes. The sessions were often quite raucous, and a so-called "healing crisis" would literally spread in waves across the room, with high emotions ranging from convulsive laughter and tears to fainting spells and catatonia, in a provocative form of group hysteria that only added to Mesmer's allure and notoriety.

Another sign of Mesmer's historical impact was his ongoing friendship with the great composer Mozart, whose first opera debuted at the

Figure 1.1—The glass armonica and its inventor, Benjamin Franklin

Mesmer estate, and a character in his opera *Cosi fan tutte* is healed with magnets. On another visit to Mesmer, Mozart first heard music played on a new instrument, the glass "armonica" (as it was then spelled), consisting of a series of progressively larger rotating glass discs that created sound much like a moistened finger circling the rim of a champagne glass.

Mesmer, himself an accomplished musician, occasionally enhanced the unearthly atmosphere at his baquet sessions with the eerie wail of his glass harmonica (have a listen at the Mozart/YouTube link to one of his final compositions, the Adagio in C Major for glass armonica, K.617).

The Science of Animal Magnetism

In 1779, Mesmer published his *Treatise on the Principles of Animal Magnetism*, which outlined the twenty-seven postulates of animal magnetism. His attempt to find a scientific explanation for the experiences he was observing was highlighted by four key propositions:

- A subtle *fluidum* connects the planets, the Earth and humanity. (This was Mesmer's term for the liquid-like energy infusing all of life).
- Illnesses arise from an unequal distribution of this fluid; restoring its equilibrium leads to the recovery of health.
- The fluid can be directed to different parts of the body, and to other people, via inner channels.
- Provoking a "healing crisis" can cure disease by releasing blockages to the flow of fluidum.

As we'll see in the next section, the similarities between this model and many non-Western forms of medicine are striking. For example, Mesmer's fluidum parallels the qi of Traditional Chinese Medicine, while his inner channels resemble the acupuncture meridians, and so on. Regardless of these historical parallels (which they were surely not aware of), the Parisian medical establishment did not look favorably on Mesmer. To counter mounting professional opposition to his work (and to his financial success), Mesmer proposed a series of demonstrations to prove the clinical value of animal magnetism (Donaldson 2005, 572–575). However, the medical establishment initially refused to even consider this option, fearing that it would unintentionally bestow validity on Mesmer and this controversial practice.

The conflict was further enflamed by a "pamphlet war" (the social media of its day) between the two sides. Eventually, in 1784, King Louis XVI appointed a Royal Commission to evaluate Mesmerism. Benjamin Franklin, then the U.S. Ambassador to France, was among the esteemed members, along with the chemist Antoine Lavoisier, the discoverer of oxygen, and the noted surgeon Joseph-Ignace Guillotin, whose eponymous device was widely used during the post-Revolution "Reign of Terror" to execute Monarchist sympathizers.*

The format Mesmer proposed for testing his therapy was, in essence, the first controlled clinical study in the history of Western medicine (Donaldson), because it aimed to assess clinical outcomes by comparing the status of a series of patients who were evaluated before and after the treatment in question. More importantly, the subjects were to be blindfolded, so they wouldn't know if they had been treated with mesmerism or a sham treatment that looked like mesmerism but was nonmagnetic in nature. We would now call this protocol a single-blind study, since the patients didn't know whether they were getting real or sham treatments.

To further minimize the role of nonspecific healing factors like hope and the power of suggestion, modern research now advocates the even more rigorous double-blind protocol, in which the doctor herself wouldn't know which treatment she is administering (e. g., whether the pill being dispensed is an active medication or a placebo sugar tablet).

This more rigid format is the modern goldstandard in clinical research because it eliminates the influence of experimenter bias, the

---

*Including several of his fellow Commission members; the judge who pronounced sentence on Lavoisier said "The Republic has no need for scientists," a sentiment that has chilling resonance in today's America.

ways a treating doctor might subtly favor one treatment and give unintended suggestions or behaviors in that direction (i.e. by tone of voice). As an aside, the use of the double-blind methodology in any interactive treatment, from psychotherapy to acupuncture, is a challenge: How can the provider be blinded to the type of treatment she is giving? Nevertheless, the Mesmer study was an important first step in the development of modern medical research protocols.

Despite these trailblazing steps, the French study had some serious limitations. One concern was that animal magnetism treatments were not to be administered by Mesmer himself but by a former student, Dr. Charles d'Eslon, who had recently been threatened with expulsion from the prestigious Society of Medicine if he continued to promote Mesmer's teachings. This professional bind created a powerful inherent source of experimenter bias: d'Eslon's career would be in jeopardy if his findings favored Mesmer, a consideration which would certainly impinge on his healing efforts. (A notorious 1999 study about energy fields, to be discussed in chapter 2, was later shown to be invalid because it ignored this very factor of experimenter bias.)

Also, the French Royal Commission chose to focus on the existence of the magnetism itself rather than on its possible clinical benefits. In other words, they opted to explore the possible mechanism of action for mesmerism, rather than first assessing whether it had any clinical impact. But which is more important, knowing whether a treatment is helpful to patients or understanding how it works? It's a research dilemma that's still alive today, as many widely prescribed medications were used for years before their mechanisms of action were discovered, with aspirin and narcotic painkillers being two famous examples.

The Commission found that many subjects indeed felt better after treatment, although they could not reliably distinguish the magnetized from the unmagnetized objects they were asked to hold or touch (including stones, glasses of water, and trees). The subjects were often unaware that, while blindfolded, they had themselves been magnetized.

The Commission thus concluded that the force called animal magnetism simply did not exist, and any apparent benefits of mesmeric treatments (and many were noted) were due to the power of imagination and suggestion—what we would now call the placebo effect—rather than to any invisible healing force. As a result of this report's extremely rapid and wide dissemination through the medical communications network (twenty thousand pamphlets were printed in one week, the 1780s equivalent of going viral) Mesmer was branded a charlatan and his reputation destroyed. He left Paris and moved to Austria, where he died in 1816 (Ellenberger 1970; Gauld 1992)

Mesmer's Legacy

Despite the dramatic collapse of Mesmer's career, his students maintained a network of mesmeric infirmaries that spread throughout Europe, and a journal of medical magnetism. *The Zoist: A Journal of Cerebral Physiology and Mesmerism* was established to report on their clinical findings. In response to this continued wave of interest in mesmerism, the French Academy of Medicine voted in 1826 to reconsider animal magnetism. Because Mesmer had died ten years earlier, the Academy's focus could not be diverted by his charismatic presence or powerful professional influence, so their replication study was thus less controversial than its predecessor. It included a demonstration of mesmeric anesthesia's efficacy for the surgical excision of a cancerous breast (Winter 1998), as well as the direct experience of magnetism by the investigating scientists themselves. In 1831, the Academy produced a much less widely publicized report that judged the phenomenon of animal magnetism to be genuine and worthy of further exploration.

Notable among the successes of mesmerism after Mesmer was its use by a prominent Scottish naval surgeon to induce full surgical anesthesia in 150 patients (Gauld 1992) (figure 1.2). No infections or deaths were reported, in contrast to the highly risky "bite-the-bullet and gulp-the-whiskey" protocols used in America's Wild West at that time.

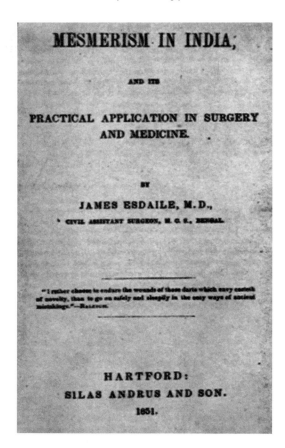

Figure 1.2—Esdaile's report (1851)

The somewhat blurry quotation from Raleigh on the front cover of the monograph *Mesmerism in India* (figure 1.2) is both eloquent and inspiring: "I rather choose to endure the wounds of those darts which envy casteth of novelty, than to go on safely and sleepily in the easy ways of ancient mistakings." Such was the courage of the post-Mesmer magnetizers.

Despite these important ongoing advances, the practice of mesmerism faded quickly following the first successful demonstration of chemical anesthesia with ether in 1854 in Boston's Massachusetts General Hospital (figure 1.3). The chief of surgery at Harvard was in attendance and famously proclaimed "Gentlemen, this is no humbug!"—inhaling ether fumes had induced full surgical-depth anesthesia in a matter of

Figure 1.3—The first successful demonstration of ether anesthesia
(Boston, 1854)

minutes, compared to the hours-long preparation that mesmeric anesthesia often required. Only minimal training was needed to use this new procedure, and so chemical anesthesia (using ether or chloroform) rapidly became the standard of care in surgery.

This breakthrough relegated the study of any potential health benefits of applied magnetism and electricity to the fringes of respectable science for almost one hundred years. In fact, the only reference to animal magnetism that has persisted in contemporary culture until recently arose from a misunderstanding. People will commonly say, for example, that they were "mesmerized" by a musical or theatrical performance, as a synonym for "entranced" or "hypnotized." However, the two processes—hypnosis and mesmerism—are actually quite different. Hypnosis uses only the spoken word to change the patient's focus of attention in order to bring about therapeutic effects, while mesmerism uses direct manipulation of the biofield to create positive outcomes. Ironically, the dangling watch chain favored by Hollywood hypnotists may actually trigger a shift in the subject's animal magnetism, as we'll see in chapter 5's discussion of EMDR.

Figure 1.4—The author reclaims the Ether Dome

So ended the story of Mesmer, apart from one historical footnote of some interest. In 1999 I was invited by the Pain Management Program of the Mass General Hospital to deliver a lecture on "Mesmer and the History of Hypnosis and Pain Management." It took place in the same surgical amphitheater where the famous ether demonstration had occurred almost 150 years earlier (Figure 1.4). I felt as though I was helping to restore Mesmer's rightful place in medicine's pantheon by speaking of his successes and insights while standing in one of conventional medicine's most sacred sites, the shrine of the Ether Dome.

WESTERN VITALISM

In retrospect, it's fair to ask what's so terrible about magnetism that it should have been mocked and shunned as a possible healing resource for so many decades after Mesmer? The problem is that it challenges the materialism at the core of Western medicine, the idea that health

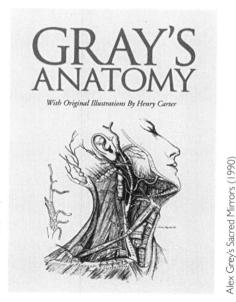

Figure 1.5—ATP

(and reality) is a product of concrete, tangible things, even if some of these are microscopic and invisible to the naked eye. If any components or forces can't be seen or measured, then they're simply not real, and up until recently that's been the excuse for dismissing subtle energies.

This materialist bias is why the term "energy" has always referred medically to the chemical reactions that occur inside a cell and propel all the functions in our body, from cell growth to muscle flexing to digestion and even thinking. These processes all proceed because of one chemical in particular—adenosine triphosphate (or ATP). This is the

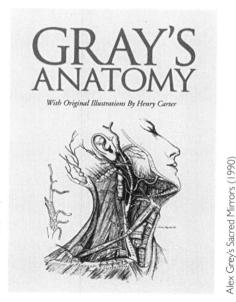
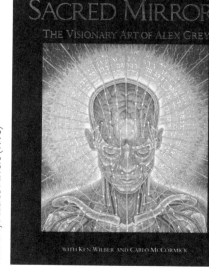

Gray's Anatomy (1858)

Alex Grey's Sacred Mirrors (1990)

Figure 1.6—Græys' Anatomies, gross and subtle.

storehouse of chemical energy, in the form of electrons, that the body manufactures to be our biochemical fuel (figure 1.5).

But Mesmer wasn't talking about ATP, and neither are any of the other energy medicine modalities being practiced today. They're talking about something that looks more like this painting and book cover by the visionary artist Alex Grey (figure 1.6).

This is what Grey sees when he looks at a human body. The nuts and bolts of muscles and bones are there, but they are interpenetrated by a series of well-defined energy structures that have been called our "subtle anatomy," to contrast with the gross anatomy studied in medical school. (And no, the classic medical textbook *Gray's Anatomy* was not written by Alex's great-great-grandfather, though Alex's book could be seen as the "revised edition" of Gray's Anatomy.) The shining lights and glowing lines in Grey's image are the key to understanding life energy and the many healing techniques of energy medicine built on this foundation. Similar pictures appear in many cultures around the world. Whether or not they were generated with the aid of psychotropic substances, their universal occurrence suggests that subtle anatomy is an inherent part of the human organism and not the hallucination of an artistic hippie.

Western scientists continued to be fascinated by this intangible energy; the following list prepares us for the complete cartography of energy medicine in chapter 5.

1. **Odic force**: This term was coined in 1845 by Prussian metallurgist and chemist Baron Karl von Reichenbach in honor of Odin, the Norse god of war (and of poets, ironically). This force was felt to be a variant of electricity and magnetism and permeated all living things. Among its many functions was the direct propulsion of the human nervous system. Sensitive people could perceive the Odic field around living things, crystals, and magnets, though only when in total darkness. Although people could be trained to emanate Odic force through their hands, mouth, and forehead, it never attained great popularity as a healing tool.

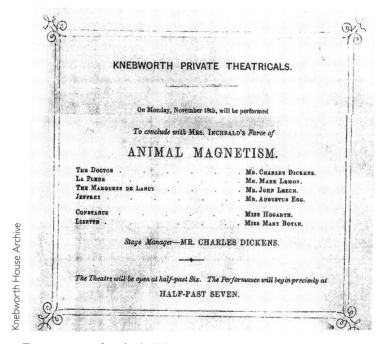

Knebworth House Archive

Figure 1.7—The playbill for Dickens's performance as Mesmer

2. **Vril**: A British science fiction novel by Baron Edward Bulwer-Lytton (1871), *The Coming Race,* coined this term for an electro-magnetic elixir that gave superhuman powers to a subterranean master race. The author was a student of mesmerism, and his country estate hosted a play about Mesmer, starring his close friend Charles Dickens as the good Dr. M. (Dickens was also the Stage Manager, and a skilled amateur mesmerist in his own right) (Kaplan 1975) (figure 1.7).

The term *vril* was later hijacked by the occult wing of the Nazi party and used to explain the supremacy of the Aryan race, as they presumed themselves to be endowed with it in excess. The Nazis formed a Vril Society that used mediumship in their attempts to gain access to advanced UFO antigravity technology. Oddly, the name of a still popular British nutritional drink made from beef extract—Bovril—is a contraction of "Bovine Vril," so-called because it promised to confer the strength of an ox on its consumers. None of my British friends were aware of the origin of this brand name or its sinister associations (Wikipedia, Bovril).

3. **Élan vital**: French philosopher Henri Bergson's (1911) theory of vital energy was dismissed by the mechanistic scientists of the day. He presumed this creative force to be the motive power behind human evolution, in contrast to Darwin's theory of natural selection by a mechanical process of elimination. His book *Creative Evolution* was awarded the Nobel Prize for Literature in 1927, though it was not a work of fiction.

4. **Entelechy**: German biologist Hans Driesch believed that the processes of cell division and differentiation were organized by an innate biological force called *entelechy*, a neologism first coined by Aristotle to denote an organism's "being-at-work." In the early 1900s, he proposed that epigenesis (the changing of an organism's developmental blueprint in response to environmental forces) was a better explanatory model than the prevailing view that preformed miniature adults existed inside each sperm cell. Now a hundred years later, Driesch is being vindicated by research showing that many mind/body approaches can actually transform one's DNA (Church 2014). If in fact an organism's genetic template can be modified by external forces, it'd be an ironic resurrection of the Soviet-era version of the Lamarckian theory of evolution (the proposition that Communist beliefs and behaviors were so powerful that they could be inherited by the next generation).

5. **Etheric energy**: Classical physicists of the eighteenth and nineteenth centuries held that light waves were carried by a subtle substance called the *luminiferous ether*, much as ocean waves are transmitted by water. This term was borrowed by the Theosophists, a leading British school of mysticism that developed a hierarchy of energies of which *etheric* was the densest. Experiments performed at the end of the nineteenth century by Scottish physicist James Clerk Maxwell are generally described

as disproving the existence of the ether. However, a recent revisionist perspective on Maxwell has shown that his famous four equations only represent a portion of his work, and his remaining equations do in fact establish a basis for the existence of ether (Rubik and Jabs 2018). This paper is an attempt to overcome overt censorship of a key scientific discovery.

6. **Libido**: The most influential manifestation of Western vitalism was as *libido,* a key tenet of Freudian psychoanalysis. This Latin word for "lust" was used by Freud to mean life force energy, and its socially acceptable management was the task faced by the ego during the process of psychological maturation. Freud spent many years unsuccessfully seeking to find a neurologic basis of libido, but ended up treating libido as a psychological force that could be directed by the mind, in the process called sublimation, into powering the higher pursuits of the intellect. Ironically, the official translation of Freud's works from German into English renders his term *die Seele* as "the mind," even though it literally means "the soul," thereby depicting the mind as a psychological rather than spiritual organ. Whether intentional or not, this careless translation strips his work of a crucial spiritual dimension (Bettleheim 1982) that imposed an artificial constraint on the field of psychiatry for decades.

7. **Libido after Freud**: His earliest disciples expanded Freudian orthodoxy in many directions, with several becoming well-known even in contemporary culture, including:
   - *Alfred Adler*—He developed the widely used terms introvert and extrovert to describe basic personality styles that determine how behavioral energy is expressed or conserved.
   - *Anna Freud*—Freud's daughter became a psychoanalyst and wrote about how ego defense mechanisms are used to regulate the energy of the libido by using such tactics as projection, suppression, displacement, and denial to protect our vulnerable sense of self (Freud 1936).

- *Carl Jung*—He wrote about the shared collective unconscious mind of humanity, a function that manifested in the form of universal myths, archetypes and even parapsychology. He begat Joseph Campbell and the breakthrough PBS series *The Story of Myth* (1988), as well Dr. Stanislav Grof's early research into the universal symbols shared by dreams, hallucinations, and psychedelic visions.

8. **Wilhelm Reich:** Also among the generation following Freud, Reich's interest in treating neuroses led to his discovery of *orgone energy,* the energy that is released during a sexual orgasm. He also created devices and therapy practices that would free up the healthy flow of this totally sexual form of life energy. One such device, the Orgone Accumulator, is shown in figure 1.8. He is a major figure in the story of subtle energy and will be discussed in more detail in chapter 2.

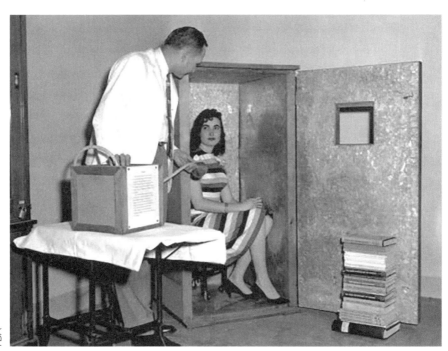

Figure 1.8—Inside an Orgone Accumulator

9. **The Orgasmatron**: Science fiction fans may remember the 1973 Woody Allen movie *Sleeper,* in which the cryogenically preserved hero is thawed out two hundred years later, only to discover a world ruled by technologic despots whose most important tool of social control was the Orgasmatron. This portable globe gave people a daily dose of the orgasmic energy that was the key to their health and survival. It was also available in a full-size cabinet model that was likely inspired by Reich's orgone accumulator. In fact, Reichian therapy had become quite popular in Hollywood in the '50s and '60s, and was publicly endorsed by such celebrities as Norman Mailer, J. D. Salinger, and Sean Connery (which may explain why women found James Bond so irresistible!).

10. **L-Fields**: Harold Saxton Burr was an anatomy professor at Yale who, from 1930–1970, studied the electrical properties of growing cells. He described patterns of electric charge changes in cells during wound healing, menstruation, cancer growth, and regeneration. He called these the fields of life, or L-fields, and considered them to be the foundation of his electrodynamic theory of development, which gave pattern to all life (Burr 1973).

11. **The Force**: Yoda, the well-known Jedi Master, lived "long ago, in a galaxy far away," but it wasn't until the 1970s that his work received wide attention on planet Earth through a series of movies, prequels, and sequels (*Star Wars*, for the uninitiated). He included training exercises involving such parapsychological techniques as levitation and psychokinesis. Simple slogans like "May the Force be with you" made this potentially esoteric concept very accessible to a mainstream audience, thereby setting the stage for a wider cultural acceptance of the reality of healing energies.

12. **Deltrons**: A very extensive scientific model of subtle energies was developed in the 1980s by the former chair of the Department of Material Sciences at Stanford University, William Tiller PhD. He envisioned a parallel universe of subtle matter that mirrors our physical world and proposed that input from those subtle realms

powered our material existence. The intermediary particle that transmitted the energetic interactions between these two realms was called the deltron, and it was the mechanism by which human intentions could be imprinted and stored in metallic devices, as his later research demonstrated (Tiller, Dibble, and Fandel 2005).

13. **Torsion fields**: Physicist Claude Swanson endeavored to develop a model that could explain all the forces of the universe, from gravity to consciousness. It was based on the concept of torsion fields—micro-vortices by which energy shifted between dimensions, like miniature versions of astronomy's black and white holes (Swanson 2011).

14. **Scalar waves, zero-point energy**: These terms have become popular in New Age and New Science circles as shorthand descriptors of the quantum physics perspective on the problem of subtle energy. Most current research on subtle energy utilizes the language of quantum fields. This model attempts to explain such anomalous phenomena as light particles interacting at a distance via quantum entanglement, the consciousness of the observer impacting the behavior of light waves, and consciousness itself not being limited to the brain or by space and time. If the entire universe is permeated with invisible energy even when our telescopes perceive only the vacuum of deep space, and if this omnipresent zero-point energy can be accessed and utilized, the world would no longer need to rely on fossil fuels and other eco-unfriendly sources of energy.

Mainstream science won't accept these ideas, but an impressively rigorous science of subtle energy already exists (see, for example, the Journal of Scientific Exploration) while popular treatments of the above theories can be found in the works of Dean Radin, Greg Braden, Nassim Haramein, Shamini Jain and others (see Resources, p. 274). Next up is the story of how Eastern perspectives of life energy finally made their way to the West.

RICHARD NIXON SAVES WESTERN MEDICINE

During my college days, I was an anti-war activist and a strong supporter of George McGovern during his ill-fated 1972 run for president against Richard Nixon, my cohort's designated "bad guy." But I'll give Nixon credit for at least one thing—he paved the way for energy medicine to enter the United States. Not intentionally, but his so-called "open door" policy to China unintentionally set the stage by enabling high-level diplomatic visits between the two countries for the first time since the post-World War II Maoist revolution. In 1971, during one of the first such visits to China by foreigners, the top reporter for the *New York Times* developed a case of acute appendicitis that required emergency surgery at Peking's Anti-Imperialist Hospital. The procedure went well, but that's not why it became an international news item—that was because the surgeons successfully treated James Reston's postoperative pain and discomfort with acupuncture (figure 1.9).

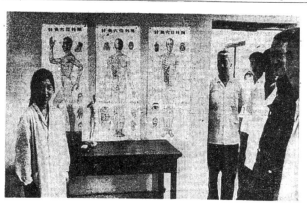

Figure 1.9—New York Times coverage of Reston's appendectomy

Reston mused that the attack of appendicitis was stress-induced, triggered when his hoped-for first-ever interview with Chairman Mao was preempted by Henry Kissinger's announcement that Nixon himself would soon be visiting China. The illness wasn't some journalistic trick to learn about the Chinese medical system from the inside, but the publicity around his positive response to acupuncture opened American eyes and minds to the possibility that balancing invisible energies could have dramatic health benefits. Mr. Reston's obituary in 1995 highlighted this event as much as his two Pulitzers and forty-five years with the *Times*.

Chinese immigrants brought acupuncture to America in the 1840s when they came to work on the railroads built for the California gold rush. However, it was still a culturally fringe practice in America through the 1950s and '60s, with acupuncturists being arrested for practicing medicine without a license as recently as 1974. But the wave of interest generated by Reston's story provided a legitimacy that might have taken decades to attain by other means. There was no going back, and Western medicine has never been the same.

Figure 1.10—The Cartesian view of pain

This wave of new ideas came just in time, because Western medicine was getting complacent, resting on its current laurels of gene therapies and organ transplants, neuromodulation and psychopharmacology. Those developments sound impressive, but how much progress had really been made? The following examples from the world of pain management clearly show why a shift in paradigms was so urgently needed.

In 1664, French philosopher René Descartes proposed that sensations of pain were carried from the peripheral injury to the brain by newly discovered tubes called nerves (figure 1.10). Three hundred and fifty years later, the world of pain management is dominated by the so-called Gate Control Theory of Pain (Melzack 1996) (figure 1.11).

Apart from the fact that color diagrams are now possible, this model is surprisingly, and depressingly, like the model proposed by Descartes. True, we now know a lot more about brain function and

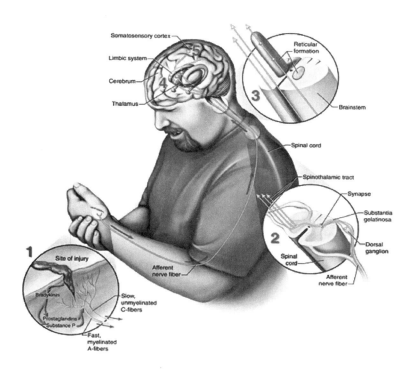

Figure 1.11—The Gate Control Theory of Pain

the complicated neuroanatomy that supports it (including the several synaptic stops on the path to the brain), but the basic mechanism remains one of neurology and physiology. The brain is also given an expanded role here as the gatekeeper for incoming pain signals, at least acknowledging a mind/body link that influences how we perceive pain.

So we should think twice before we chuckle at the simplistic model of brain function espoused by eighteenth century phrenologists (figure 1.12). For these "doctors," every bump and furrow in the scalp revealed the psychological function of the brain region underneath it (caution, veneration, humor, etc.). Personality types could supposedly be predicted from the contours of the skull. For example, a case series from the 1840s analyzed plaster casts of the skulls of executed murderers, which purportedly showed a unique phrenological pattern (Gauld 1992).

And yet, how different is that fanciful brain map from a modern diagram of the brain's functional neuroanatomy? Speech perception is in the left inferior temporal lobe, decision-making is in the anterolateral frontal cortex, spatial perception is in the parietal lobe, and so on (figure 1.13). Each function still originates in a designated brain area.

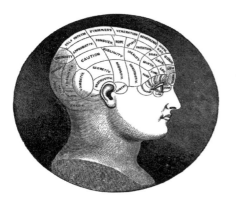

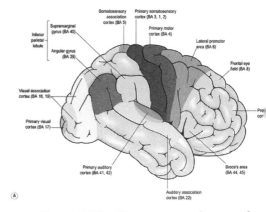

Figure 1.12—Phrenology's map of brain function

Figure 1.13—Neuroanatomy's map of brain function

If these mechanistic models are all we've got to show for three centuries of scientific progress, maybe we ought to be looking elsewhere. It's easy to snicker at the seven blind men poking at the elephant, each thinking they've discovered the true essence of this unusual and foreign creature, but conventional medicine does the same thing. Each specialist and sub-specialist is convinced that their niche is the true cause of illness, rather than just another facet of the mysterious creature known as "health."

A better comparison might be to the drunken man who is on his hands and knees in the parking lot trying to find his car keys. A passer-by asks him why he keeps looking in that one spot under the streetlight, to which he replies, "That's where the light is." If medicine continues to look only at what is illuminated by the five physical senses (and their technological extensions), they'll never find the keys to health either. Every other healing tradition on the planet has looked under many streetlights, even using their internal flashlights to detect subtle signals and invisible forces, all in the service of better understanding health and illness. So let's shine a light on how the rest of the world views health and illness, starting with the terms that these non-Western traditions use for life energy. Despite their widely varying origins, many of these terms have begun to creep into our everyday vocabulary including:

Qi/Chi—Traditional Chinese Medicine (TCM): the term is widely used in Chinese martial arts such as qi gong (energy work), exercises to cultivate energy flow, while *de qi* describes the distinctive sensation generated when an acupuncture needle hits the correct spot and "obtains qi." Surprisingly, the well-known term *tai chi* is not related to *qi.* It's more accurately spelled (and pronounced) *tai ji,* meaning "supreme ultimate path," and does not specifically refer to energy.

Ki—Japanese Zen: Ki is related to the Chinese term *chi* by etymology and is embedded in the names of the popular energy therapy *Reiki* (meaning universal healing energy) and the martial arts practice *aikido* (the way—*do/tao*—of harmonious spirit).

Prana—Ayurveda (India): because yogic practitioners understand that manipulating the breath is the most direct way to manipulate energy (prana*)*, the yogic term for breathwork is *pranayama.* The yoga texts also described an entire energy body, called the *pranayama kosha* (the sheath of prana), that is the template for the physical body (known as the food sheath, because being literally constructed from food it is another important source of prana).

Ruach—Kabbalistic Judaism: the mystical branch of Judaism distinguishes this basic physical life force, *ruach*, from soul energy, *nefesh.*

Orenda—Iroquois: the spiritual force that flows through all living beings and is harnessed when man lives in harmony with nature. *Orenda* is best evoked through prayer, song, dance and sacred ceremony.

Wakan Tanka—Lakota/Sioux: meaning "great spirit," *wakan tanka* is the sacred or divine power that resides in everything. In a striking example of cultural appropriation, many Americans know this term because of the Tonka toy company, whose metal trucks were must-haves for millions of American boys in the sixties and seventies. The name was chosen because the Minnesota-based company also aspired to be "great" like the term's translation. Unfortunately the only thing that was great was their debt load, and they were bought out by Hasbro in 1991.

Huna—Hawai'ian shamanism: A kahuna was an expert in any field, possessing inner knowledge or or secret wisdom. The term "Big Kahuna" entered American slang with the 1959 surfer movie *Gidget*, when Cliff Robertson was awarded that title for his L.A. surfing skills (another example of cultural appropriation at its insidious best).

Barakah—Islamic mysticism (Sufism): the blessed power that originates in Allah and permeates all life. In another interesting bit of comparative etymology, this word is cognate with the Hebrew verb "to bless"—*baruch*—a poignant example of how modern-day cultural and political antagonists often share common origins.

Manitou—Algonquian (Ojibwe/Chippewa): the spiritual life force that is present everywhere and used by medicine men for healing (Mavor 1989). The Canadian provincial name Manitoba literally means the straits of *manitou.*

Orisha—Yoruba (West Africa): a psychological construct and an archetypal energetic entity emanating from the Supreme Being.

Mojo—Gullah/Creole: from the West African *moco'o*, meaning "shaman." It's commonly used now thanks to African American slang, where it refers to magic or magical ability.

And finally, two terms from the Western European mystical tradition:

Pneuma—Classical Greece: meaning spirit, energy and breath. Many medical terms relating to the lungs such as pneumonia come from this root, reflecting the clear breath/energy link known to the Greeks.

Spirit—Latin/Middle English: a multilayered term whose root *spiro* literally translates to "I breathe," with cognates and derivatives ranging from *spir*ituality and the Holy *Spir*it, to ex*spir*ation (of breath and of life) in*spirat*ion (of breath and wisdom), con*spir*acy (literally "breathing together"), and even in reference to alcoholic drinks or *spir*its (so-called because they were thought to open the door to evil entities.

This is only a partial list (as over one hundred names are recognized by some sources), but it illustrates how various communities perceive the world very differently than conventional physicians. Unsurprisingly, subtle energy is the key factor, the source of light that our distracted and mis-focused Western doctors have been ignoring as they look for the car key in the parking lot. This perspective affects how symptoms are treated, with pain management providing a clear contrast. As we will see in chapter 6, the Eastern approach doesn't focus on how injured tissue may trigger nerve impulses, but instead focuses on blocked energy flow as the cause of discomfort. This blockage is then treated with exercises such as yoga and tai chi, acupuncture, massage, and herbs to unblock pathways and restore a balanced flow of energy.

In chapter 6's discussion of phantom limb pain, the energy model will shine a new light on mechanisms of pain that go far beyond the gate control theory. By understanding the root of pain more deeply we can envision new therapies and promote healing at an energetic level. And as it is with pain management, so too it is with other symptoms and diseases. In summary, it looks like Kipling was wrong: East may be East and West may be West, but the twain have most definitely met, and the world will be the better for their ongoing dialogue.

2
Modern Resistance
Four Case Studies

Science progresses one funeral at a time.
THOMAS KUHN, PARAPHRASING MAX PLANCK
THE NATURE OF SCIENTIFIC PARADIGMS, 1972

German physicist Max Planck was one of the pioneers in the field of quantum physics, and as such was very familiar with the ways by which the scientific establishment could resist powerful new ideas. The complete version of his famous quote is:

A new scientific truth does not triumph by convincing its opponents and making them see the light, but rather because its opponents eventually die, and a new generation grows up that is familiar with it (Planck 1968).

We're now in the middle of just such a changing of the medical guard, and it's informative to look at several examples from the history of energy medicine (EM) science that show how opponents don't become convinced, they just fade away. We'll look at four examples from the last sixty-five years in which various forms of energy medicine have been suppressed. The parallels to Mesmer's story from two hundred years ago are striking because the key factors still operate today:

- The innovator's personal path of inspiration and discovery despite challenges
- The conservative and dogmatic nature of conventional medicine as a guild
- The financial and political pressures that hinder the development of new medical treatments
- The role of the Establishment in controlling the dissemination of information
- The way that historic legacies are shaped by The Powers That Be (TPTB)

CASE STUDY
. . . . . .
THE FDA IMPRISONS WILHELM REICH

As mentioned earlier, the vitalist thread in Western medicine was maintained, and strengthened, by one of Sigmund Freud's most misunderstood disciples, Austrian psychoanalyst Wilhelm Reich. Much as Freud did with libido, Reich initially tried to find a physiologic explanation for orgone energy, the sexualized form of life force energy he thought was the key to health. He used the scientific language of the day, publishing such papers as "The Orgasm as an Electrophysiological Discharge" (1934), but his direct work with patients evolved into an entirely new form of psychotherapy, Bioenergetics, that went beyond simple talk therapy. He encouraged his patients to move their bodies rhythmically in ways that would release their "character armor"—the tensed muscles of the arms, legs, and trunk that were supposedly trapping the orgone energy and engendering neurotic behavior patterns.

In 1930s Germany, his sexual reform movement (for which he coined the term "the sexual revolution") was seen as a Jewish plot to undermine European society, so he emigrated to America in 1939 to escape fascist censorship and death camps. In New

York City, he was able to work unencumbered. He refined his theory of the orgasm as the cure for emotional and physical illnesses and pursued the experimental treatment of cancer patients with his orgone accumulator—a box made of alternating layers of wood and metal that allegedly collected and focused this healing energy (figure1.8).

Reich even attempted to manipulate the weather with his "cloudbuster," an artillery-like adaptation of the orgone boxes. He wrote prolifically, penning eleven books (including *The Function of the Orgasm, The Sexual Revolution, The Cancer Biopathy,* and *The Mass Psychology of Fascism*), and many articles that were published in mainstream psychiatric journals. Despite the controversial nature of his work, he only came to the U.S. government's attention in 1947, following a series of critical articles in the popular press (*The New Republic* and *Harper's*) that called him the leader of "a new cult of sex and anarchy."

Investigators from the FDA—a branch of the federal government with enforcement abilities that come from its centralized political power—believed that Reich's health claims were fraudulent. When he refused to cooperate with an unannounced on-site visit

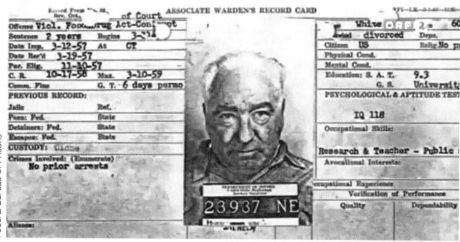

Figure 2.1—Reich's prison mug shot

in 1954, the FDA sought a permanent injunction against the interstate shipment of his accumulators. Reich refused to appear for his hearing and was charged with contempt of court. His accumulators were confiscated and destroyed, and six tons of his books and journals were destroyed in the only federally sanctioned book burning in American history (Laing 2001). After an FDA sting operation entrapped a colleague shipping accumulator parts across state lines, Reich was convicted of those contempt charges and sent to a federal penitentiary. He died of a heart attack in 1957, one week before his parole hearing (figure 2.1).

Despite facing drastic censorship on par with Ray Bradbury's sci-fi classic *Fahrenheit 451*, his influence has continued. As mentioned in chapter 1, many celebrities of the 1960s owned accumulators, and Woody Allen likely invented the "orgasmatron" as a parody of Reich's ongoing influence in Hollywood's hip circles. Thanks to Reich, the field of psychotherapy expanded beyond simple talk therapy, and the body-centered therapy movement that began with Reichian bioenergetics has expanded to include less sexualized approaches, such as:

- **Gestalt therapy**—popularized by Fritz Perls and the Human Potential Movement at California's famous Esalen Institute, Perls investigated how body language and posture communicate emotional truths that words can cover up, calling his approach Gestalt therapy (Perls 1969).
- **Focusing**—troublesome emotions are not just talked about, but are felt in the body and regarded as metaphorical visitors to be befriended rather than repressed or ignored, in order that their messages can be received and heeded (Klagsbrun, 2001, 115–30).
- **Somatic Experiencing**—progressively experiencing and releasing the internal sensations that accompanied past trauma but that remain stored in the body and are seen as the key to healing (Levine 1997).

- **Meridian-Based Therapies**—techniques like Emotional Freedom Technique (EFT) and Thought Field Therapy (TFT) stimulate the acupuncture meridians by tapping on the body to bring about emotional release and energetic rebalancing.

So while the U.S. government may have literally destroyed Reich the man, his ideas live on and continue to flourish.

CASE STUDY
• • • • • •
THE AMA DEBUNKS THERAPEUTIC TOUCH

Compared to the FDA, the American Medical Association (AMA) does not have direct governmental power but wields vast influence as the guild and lobbying arm for physicians nationally. Ever since the Carnegie and Rockefeller-sponsored Flexner Report in 1910 established parameters for the growth of modern conventional medicine as we know it, natural healing modalities have been censured, med school curricula have become more pharmaceutically based, and the AMA has become the nation's most powerful alliance of clinicians. Of note, the AMA did lose a major lawsuit for restraint of trade filed by a rival chiropractor guild in the 1970s, *Wilks v. AMA*. The AMA's case was undermined by the release of insider information obtained, oddly enough, by another of conventional medicine's adversaries—L. Ron Hubbard and the Scientologists. (Weeks 2020).

In general, the AMA's influence on medical research and clinical practice was magnified when *The Journal of the American Medical Association* (JAMA) became the profession's gold standard and the gatekeeper of conventional medical orthodoxy. Given their rigorous editorial standards, low rate of manuscript acceptance, and large international readership, the high point of many a doctor's research career was having an article published in *JAMA*.

So, it was a shock for energy medicine practitioners to find that the feature article in the April 1, 1998, issue of *JAMA* was a study entitled "A Close Look at Therapeutic Touch" (Rosa 1998). This study tested a popular form of energy therapy called Therapeutic Touch (TT), in which trained practitioners detect energy blockages with the palms of their hands and then smooth out perceived irregularities in their patients' energy fields. In this controlled study, nurse practitioners of TT were blindfolded (making it literally a single blind study) and asked which of their hands, (i.e., which part of their energy field) was being approached by the experimenter's hand (figure 2.2). It was unprecedented for *JAMA* to focus its esteemed attention on the questionable world of energy medicine, and also intriguing since the study seemed to be a valid test of one key tenet of the practice: does a biofield actually surround the human body, and can it be detected (and presumably manipulated) by trained practitioners?

Figure 2.2—Emily Rosa and her experimental setup
Illustration by Rosi Fatah

Since the nurses chose correctly at less than the 50/50 success rate of pure chance, the article concluded that life energy simply does not exist. Then, in a highly unusual move, *JAMA's* Editor-in-Chief also added a black box warning of the type that is usually reserved for health emergencies:

> Practitioners should disclose these results to patients, third-party payers should question whether they should pay for this procedure, and patients should save their money unless or until additional honest experimentation demonstrates an actual effect.

Apart from besmirching the reputation of TT researchers by implying that they were dishonest, this edict against using TT for patient care defied logic because the study didn't even look at clinical outcomes. It only focused on one possible mechanism of TT, much as the French Royal Commission didn't explore the therapeutic impact of mesmerism, but only focused on the existence of animal magnetism itself. Mechanism of action is ultimately only a secondary consideration, as many effective therapies are used in modern medicine even though we don't know how they work (ether anesthesia, for example).

Nevertheless, on the same day that the *JAMA* article appeared (April Fools' Day, as luck would have it), the story was making headlines nationally, including a front-page article by *The New York Times* lead science reporter (Kolata 1998), along with multiple network TV appearances by the lead authors. It was an updated but still pre-internet version of Mesmer's pamphlet war (the internet will be rearing its ugly head in Case Studies #3 and #4). And that's where the story got even more interesting, as it became clear that the article was actually a write-up of the elementary school science fair project of eleven-year-old Emily Rosa, one of the paper's lead authors. She had been inspired by a

TV documentary on TT and was guided by her parents, who were also coauthors and members of the National Coalition Against Health Fraud. The final lead author, Dr. Steven Barrett, was the founder of an organization called Quackwatch Inc., a group whose name clearly reveals its non-neutral opinions with respect to alternative medicine (as it was then called). Researchers are of course allowed to have professional and political affiliations, but such clear partisan ties raised eyebrows when the potential for bias in this study was not acknowledged by *JAMA*.

When independent researchers analyzed the study more closely postpublication, several flaws in methodology were uncovered that were serious enough to question the results. Most importantly, the problem of experimenter bias was never addressed, since Emily was most likely a TT skeptic (that seems to be a reasonable assumption, considering her parents' open bias), had not been trained in TT, and would not be expected to have any particular ability to project energy. Since the mind affects the flow of energy. Emily might have actually been unconsciously retracting her own energy field, thereby making its detection by the TT nurses *more* difficult. In fact, an earlier study of energy sensitivity using a very similar blinding technique but done with trained energy healers in both roles (as senders and detectors) found a much higher rate of successful detection than the Rosa study (65% versus 44%), a statistically significant success rate (Schwartz 1995). Quite tellingly, this study was not even cited in the Rosa paper's extensive list of references, bypassing the time-honored tradition of addressing conflicting data in scientific papers.

In fact, the article triggered a record number of letters to the editor, and to its credit, JAMA printed many of these critiques in a subsequent issue (Ireland, 1998). Among the numerous responses was one rebuttal that attempted to "un-debunk" the Rosa article (Leskowitz, 1998). However, the damage was done: the journal had lent its imprimatur to a flawed study, and the coordinated media push ensured that Americans developed a negative view of TT and energy therapies. So just as Mesmer was outflanked by the barrage of fliers that were quickly distributed throughout the streets of Paris, this concerted media campaign was also an effective tool for disarming another threat to the medical powers that be. Fortunately, the TT community has continued an active research program, and as we'll see in chapter 5, have used some very clever experimental protocols to eliminate the influence of nonspecific factors to show TT's efficacy (experimenter bias, the power of suggestion).

CASE STUDY
• • • • • •
WIKIPEDIA SLANDERS ENERGY PSYCHOLOGY

The first two examples of the suppression of EM (energy medicine—the umbrella term for treatment modalities that explicitly use life energy in their healing processes), suppression happened in the pre-internet era, and it would be comforting to think that the introduction of social media opened the floodgates for the free and open flow of information in scientific and political arenas, leaving the era of de facto censorship in its wake. However, censorship is now even more pervasive, and subtle to the point of being invisible to the untrained eye. The process is rarely as blatant as the "fake news" alleged by former president Donald Trump, but the accusation has become a widely used bludgeon to dismiss alternate political opinions in the hope that they would not be spread far and wide. Practitioners of energy-based medicine have seen how social media giants have become society's information gatekeepers, though with indirect, and therefore more dangerous, techniques of disinformation and algorithm manipulation. Here is a telling vignette involving Wikipedia.

Wikipedia is the modern incarnation of the *Encyclopedia Britannica*, the one-stop repository of all the information anyone might ever need about any subject whatsoever. It is widely consulted by the lay public for medical information because its articles are continuously updated and fine-tuned by its readers, who are anonymous members of the general public and are presumed to be well informed. Content is also moderated by an active team of Wikipedia staff sub-editors (also anonymous). As the saying goes, Wikipedia is like the Bible. No one knows who wrote it, but we all believe it blindly. And for a discussion of straightforward medical issues like the health benefits of exercise or the surgical treatment of heart disease, Wikipedia does a good job. The story gets murkier, though, when using the website to research the more controversial field of "alternative medicine" (Wikipedia's favored term, one long ago abandoned by integrative health practitioners due to its judgmental tone and dismissive implication). Articles on holistic or integrative medical topics are peppered with terms like "pseudoscience" and "unproven," and include links to websites and blogs such as Quackwatch Inc. and the National Council against Health Care Fraud Inc. (this despite Wikipedia's own stated criterion of using only peer-reviewed articles rather than personal blog posts).

Against this background comes the particular EM story of Energy Psychology. I'll note that one of the most effective techniques I've encountered in my work with chronic pain patients is the acupressure-based form of psychotherapy called Emotional Freedom Techniques (EFT, or tapping). Despite a very extensive research base (Over one hundred clinical trials, fifty randomized controlled trials, and six meta-analyses) showing its efficacy for a wide range of disorders, it has been dismissed by Wikipedia as "pseudoscience." In response to the Wikipedia entry to that effect, several members of the professional organization of EFT practitioners—the Association for Comprehensive Energy Psychology (ACEP)—attempted to update the entry in accordance with Wikipedia's editing and posting guidelines. However, ACEP found that their updates were deleted within a matter of days, sometimes even hours, even though they were adding exactly the type of research data that Wikipedia supposedly sought.

To more directly address this apparent bias against a proven therapy, ACEP took the unusual step of creating an online survey through Change.org that urged the publisher of Wikipedia—Jimmy Wales—to update their entry on EFT to include recent research that validated its efficacy. In 2016, over 10,000 licensed clinicians signed the petition and much to our surprise (I am a member of ACEP), Mr. Wales responded directly. However, his response was, to put it mildly, not constructive. After stating that Wikipedia would of course post the results of valid science, he said:

> No, you have to be kidding me. Every single person who signed this petition needs to go back to check their premises and think harder about what it means to be honest, factual, truthful. . . . If you can get your work published in respectable scientific journals—that is to say, if you can produce evidence through replicable scientific experiments, then Wikipedia will cover it appropriately. What we won't do is pretend that the work of lunatic charlatans is the equivalent of 'true scientific discourse.' It isn't.

While "The Lunatic Charlatans" might make a catchy name for a punk rock band, its use by Mr. Wales was unprofessional, inappropriate, and possibly even legally actionable as libel (a path not pursued by ACEP). Wikipedia's treatment of EFT hasn't changed noticeably at the time of this writing (2023), despite continued growth of the EFT research base. Wikipedia still cites the non-peer-reviewed *Skeptical Enquirer* website, the term "pseudoscience" persists, and at least one blatantly false charge is highlighted in the article's summary: EFT is said to show "no benefits as a therapy beyond the placebo effect," a charge directly addressed and disproven by extensive research. While skeptical inquiry is central to the mission of science, this response demonstrates a lack of neutrality and objectivity in some of its self-styled practitioners.

The full ACEP/Wikipedia story fits into a larger pattern of pseudo-skeptics attempting to debunk holistic therapies by posing as neutral observers when they are in fact highly partisan (Leskowitz 2014c). For example, the founder of Quackwatch Inc., Dr. Steven Barrett, also a coauthor of the TT/*JAMA* study, has been the target of multiple legal challenges and has had his professional integrity and competency questioned by summary statements issued by judges in several adverse court rulings (Gale 2019). An important website called "Skeptical About Skeptics" outlines the system used by Wikipedia to monitor all incoming revisions and updates on controversial topics, and then assign a rapid response team of anonymous editors to scrub nonconforming updates quickly and efficiently. The common denominator among all the targeted therapies—EFT, Reiki, and even acupuncture—is that they are energy-based therapies. So Wikipedia effectively serves as a self-appointed and unmonitored gatekeeper, one of whose primary functions is to ensure that energy medicine in any of its forms isn't accepted as legitimate by the general population.

CASE STUDY
• • • • • •

SOCIAL MEDIA CENSORS
"THE DISINFORMATION DOZEN"

Most recently, many social media giants have begun to push back against popular natural health websites, even more noticeably with the onset of Covid and the spread of holistic approaches to its prevention and treatment. One particularly sinister pre-Covid move was the June 2019 behind-the-scenes rollout of Google's new search algorithm. Within one week, such heavily trafficked integrative medicine websites as Dr. Joseph Mercola's website were seeing a 95% drop in visits, with Google searches for

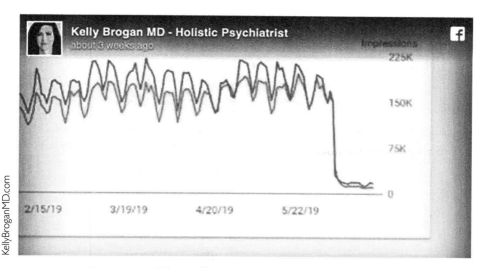

Figure 2.3—The traffic pattern at Dr. Brogan's website

natural health topics now re-routing queries to mainstream health sites and active critics of integrative medicine. Figure 2.3 shows how the traffic stats for the website of holistic psychiatrist Kelly Brogan abruptly dropped from 200,000 daily hits to 10,000, literally overnight.

Since web traffic is the lifeblood of these sites, the new algorithm became a virtual death sentence (in both senses of the word "virtual"). However, independent search engines like Bing and DuckDuckGo emerged as popular alternatives to Google because, in addition to providing clear access to controversial sites, they also afforded the sort of privacy protection that has been a growing concern of Google and Facebook users. This pattern of de facto censorship by algorithm has become even clearer with respect to criticisms of the Covid vaccine program. Overt suppression of dissident "anti-vaxxer" views via "deplatforming" has become the norm on Facebook, YouTube, Twitter, and Google, even when the presentations feature highly credentialed health care professionals.

Here's the response of one respected investigative journalist to Amazon's refusal to sell his book, a critique of the established medical and public health POV:

> I have experienced similar censorship by Medium, Mailchimp, Patreon, and YouTube. The pattern is always the same: no notice given, no contravening content specified, no broken rule listed. Just my content gone in an instant! There is never any opportunity to resolve the issue. Nobody accountable is ever named (Geddes 2021).

Shockingly, an article entitled "Spy Agencies Threaten to 'Take Out' Mercola" (2020) documents that the website of a prominent holistic physician—Joseph Mercola—has been labeled a national security threat by British and American intelligence agencies, as they collaborate to eliminate "anti-vaccine propaganda" from public discussion by using sophisticated cyberwarfare tools to detect and block such content. It's a tight squeeze, because corporate consolidation has given these unregulated social media giants a near monopoly in deciding what information should be included within the boundaries of our national conversation, decisions that had previously been made by bipartisan government agencies like the Federal Communications Commission (FCC). Unless these private corporations are broken up or regulated as public utilities, America is in the unusual position of having Donald Trump (banned from Twitter) and environmental activist Robert F. Kennedy, Jr. (banned from Facebook) sharing the same fate of being marginalized despite being at the two ends of the political spectrum. And if this is the case with politics, we should be even more concerned with restricted access to medical information and the public's right to self-education. In this vein, a lawsuit has been brought against the legacy media on First Amendment grounds that free speech is being suppressed by this sort of scientific censorship (GreenMedInfo 2023).

This sort of suppression has become easier to institute and manage now that some of the most prominent targets have been conveniently lumped together under the catchy rubric of "The Disinformation Dozen." The so-called Center for Countering Digital Hate (CCDH) reports that they have identified these twelve online content providers as contributing 73% of the disinformation that appears on social media, a statistic that was widely and rapidly covered in the mainstream media when it was released in late 2022. However, their criteria for determining what qualifies as disinformation is never specified (though criticism of the Covid vaccine and mask mandates have been common targets), and the origin story of the CCDH is quite murky. CCDH is described on its website as a British nongovernmental organization (NGO) funded by "philanthropic trusts" that are never identified, so it is impossible to tell whose interests it might represent. Plus, their widely reported interpretation of the data is at great variance with other examinations of online posting and suppression rates; one independent follow-up analysis by Facebook's VP of Content Policy revealed that, contrary to the CCDH claims, less than 0.1 percent of total online Covid-related content has originated from those twelve sources (Bickert 2021).

Furthermore, there is a significant disconnect between what these social media giants are trying to suppress and what the American public is actually able to access. The most striking example is the bestselling book by Robert F. Kennedy Jr., *The Real Anthony Fauci*, released in November 2021. This exposé of financial links between the pharmaceutical companies and federal agencies like the FDA and the Centers for Disease Control (CDC) during the Covid pandemic sold over one million copies in its first four months, despite its subject matter and author being prime targets of the de facto censorship described above. There has been no mainstream media coverage of the book's factual content or mention of its bestseller status, apart from the usual buzzwords like "conspiracy theory" and "anti-vaxxer." A typical recent opinion piece was entitled "RFK Jr. is the Dumbest Kennedy" (Callahan 2021) and the *New York Times* has even refused to carry ads for the book (Richardson 2021), though it did place the book on its weekly bestseller lists. So, despite the best efforts of Twitter, Google, Facebook, Wikipedia, Instagram, and YouTube, the emerging array of alternative sites and distribution platforms have been able to respond to the ever-present demand for alternative narratives. Ironically, Robert F. Kennedy Jr.'s book remained on the Amazon Top Ten bestseller list for months after its release, so at least one media giant seems willing to prioritize profits over politics.

In summary: *"Plus ça change, plus c'est la même chose"* (the more things change, the more things stay the same). Science does not exist apart from politics, and scientific revolutions are every bit as threatening to The Powers That Be as political revolutions. However, the walls of protection surrounding the Western medical edifice are clearly crumbling, and the structure itself is being rebuilt to accommodate the new paradigm of energy medicine. Chapter 3 will provide an inside view of that process at work.

3
Bringing Energy Medicine into Academia

My Story

I would like to be sure that we do not develop, promote or implement non-traditional treatments. . . or any reference to the magnetism of the body.
We cannot be the leader in non-traditional medicine.

MEMO FROM HOSPITAL CEO TO STAFF,
APRIL 1996

To illustrate the process by which a new paradigm—energy medicine—makes its way into the medical mainstream, I'm going to use my own experience as a case study. The story of how I developed my interests and skills in EM, and how my hospital responded to the growing evidence for these treatments, will serve as an exemplar of a process that was (and still is) happening to doctors and hospitals across America and throughout the world. This story will illustrate the institutional and interpersonal barriers against transformation, and the value of patience. The process is still unfolding to be sure, but the medical culture has already reached the tipping point, and there's no going back.

PRE-1990s
My Background

I was brought up by a loving but secular family who worshipped at the altar of science. I vividly remember my awe at seeing all the test tubes in my father's immunology research lab at the Massachusetts General Hospital. (Ironically, his office was just down the hall from the Ether Dome.) In high school, I was good at math and sciences, and felt sure that these disciplines had all the answers, even though I wasn't asking any big questions at the time. My first forays into understanding the nature of the mind came in college, in the form of psychopharmacology—theoretical and applied. The theoretical part was a class taught by Amherst College's first neuroscientist and was a paean to our new and growing scientific understanding of brain function and the chemical messengers known as neurotransmitters. But we never really considered where thoughts came from or what force animated our bodies. Consciousness was an artifact whose understanding was kicked down the road until it could be re-examined at some undefined future date, presumably after we had broken down the components into small enough parts.

At the same time college was, for me and millions of others in my Boomer cohort, an opportunity to directly experience altered states of consciousness (ASC). Unlike former President Clinton, I inhaled—and my explorations of applied psychopharmacology via marijuana helped me realize that my mind could function in more than one mode. It also opened my eyes to such diverse and popular writings as Carlos Castaneda's books on shamanism in Mexico, Andrew Weil's writings on the power of expectation in *The Natural Mind*, and even Jane Roberts's *Seth Material* (which introduced the phrase "You create your own reality"). Interesting stuff, but on a totally different track than medicine, and with almost no overlap between the two worlds.

It was a lot of information to digest in addition to the social upheaval of the early seventies that included, for me, an anti-war march on Washington and a mass arrest at a nearby air force base. Unable to gain a clear focus or sense of life purpose, I decided to postpone medical school and instead traveled to Israel to work and study on a communal farm, a kibbutz. There I was lucky enough to meet a beguiling British school teacher and fall in love (a state of mind which quickly became my favorite ASC). We married and went on a six-month honeymoon, traveling across Europe, India, Nepal, Iran, Afghanistan, and Kashmir. It wasn't a spiritual quest or a chance to study with meditation gurus, but it showed me that other systems of medicine have their own cultural validity. Was it possible that a Tibetan herbal preparation that looked like rabbit droppings could alleviate a bad case of backpacker's shoulder? How could taking someone's pulse help to diagnose where the body was in pain? And how could simple eye contact with a supposed spiritual master lead to healing? These questions guided my entire career.

Medical School

This trip also set the stage for some serious culture shock when we came back to America. Fortunately, during the summer of 1975, before starting at University of Massachusetts Medical School (UMMS), I took an important bridge step in my evolution by learning how to meditate. TM (Transcendental Meditation) was all the rage after the Beatles signed on with the Maharishi, so it was the natural choice. The value of this practice was two-fold: I came to see that I wasn't defined by my thoughts, and I increased the signal-to-noise ratio in my conscious awareness (less static from a random blitz of thoughts, more time in quiet awareness). Meditation's drugless ASC of mental clarity felt so good that I became convinced of TM's health benefits. During the Q&A after every medical school lecture, I could be counted on to ask whether meditation would help that particular disease of the day. But this was back in the 1970s, when there was a total of approximately three clinical studies on the health benefits of meditation, so I quickly learned to put a lid on it. Enthusiasm without evidence didn't cut it in the world of academic medicine.

In fact, the only academic mention of stress and its impact on health I can remember came in a microbiology class (a basic science course during the first two preclinical years) before we'd had any direct patient contact. The professor was talking about viral diseases like herpes (oral) and admitted that he suffered an outbreak of cold sores every time his mother-in-law visited. He told it as a joke, but its deep truth stuck with me: emotions affect immunity. In this case it was for worse, but presumably it could be for better as well. It was a theme I was to follow in my later psychiatry training, and which became even more relevant with the arrival of the Covid-19 pandemic.

I had my first experience of sensing energies during medical school while taking a course in kundalini yoga (not at the med school). The instructor—white robe, turban, beard—stood in front of his seated class and led us in a breathing exercise designed to increase our energy levels. This "breath of fire" was a form of pranayama, the yogic breath exercises to generate prana, life energy. When I tried to get up off the floor at the end of the session, I lost my balance. As I stumbled forward it felt like I bumped into my teacher and bounced off him, but I knew cognitively that this was impossible because we were actually several feet apart. As I can understand it now, I was bouncing off the outer edge of his very powerful biofield (the energy balloon that he had just inflated via breathwork) to the point where it was tangible and palpable to me. I didn't have words yet to explain what had happened, but that experience of sensing energy stayed with me over the years.

In my senior year of medical school I was given permission to spend my elective month with local alternative medicine practitioners. I had my first acupuncture treatment, and when Dr. Yvonne Chen needled my hand I felt a tingling that traveled up my arm and through my shoulder in a very clear and distinctively odd zig-zag pattern. I hadn't yet learned where the acupuncture meridian pathways lay, but she showed me the chart where the small intestine meridian traveled—exactly where I had noticed those prickling sensations. This clear sensory experience was an important step in validating the reality of acupuncture energy for me, despite its subjectivity, but my UMMS professors weren't buying it. For example, my anatomy professor maintained that the reason acupuncture worked so well in China was that the people there were brainwashed by Chairman Mao into a form of mass hypnosis that had nothing to do with energy and healing. Fortunately, research demonstrating acupuncture's clinical benefits has moved well beyond this colonialist and not-a-little-bit racist point of view.

I missed an opportunity to help bridge the East/West gap during my third year internal medicine rotation at Worcester City Hospital, when the Chief of Rheumatology showed us a diagram of a new syndrome he was exploring. His patients developed general fatigue plus muscle soreness around their body, but to me his "fibrositis" map looked for all the world like the acupuncture charts in Dr. Chen's office. I was too self-conscious to mention this unusual similarity to him because I wasn't yet ready to come out of the energy closet in the real world of hospitals. Fortunately, other clinicians went on to document the similarities between these tender spots of fibromyalgia (as it's now called) and acupuncture points, and to the trigger points seen in myofascial pain (Travell 2018).

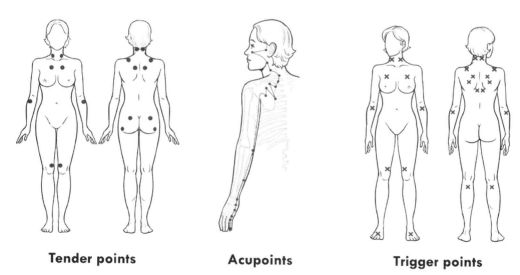

Tender points   **Acupoints**   **Trigger points**

Figure 3.1a: Fibromyalgia tender points; 3.1b Acupuncture meridian and points; 3.1c Myofascial trigger points.
Illustration by Rosi Fatah

Yet that insight, even though I suppressed it, was another key experience that supported my growing awareness that something important was going on with this ancient therapy.

I was fortunate to have a mentor in med school, a professor who was a cell biologist by day and a consciousness explorer after hours. I will always be grateful for Dr. Earl Ettienne's wisdom and guidance in introducing me to the wider world of health and healing. He saved me much angst by advising me to temper my enthusiasm for these new and as-yet-undocumented therapies by choosing my words carefully. For example, if I started speaking to my physiology professor about auras he'd just tune me out, but if I talked about the electromagnetic field surrounding the body he'd be all ears. So I became bilingual, learning how to translate woo-woo energy concepts into scientific terminology that my colleagues could relate to.

Earl also introduced me to some key players in the emerging Boston holistic medical scene, including the staff at Interface (Boston's version of Esalen) and UMMS's very own Jon Kabat-Zinn. Dr. Kabat-Zinn is now the godfather of mindfulness, but in the late '70s he was a full-time cell biology researcher who gave some lunchtime presentations to staff about his hobby—mindfulness meditation. The talks generated so much interest that he began to wonder about the clinical effects of teaching meditation to patients.

Unfortunately, the physical layout of UMMS's two-building campus made it hard for him to connect with possible partners to set up clinical trials. His lab in the basic science building was completely separated from the hospital and the outpatient clinics, with only a single corridor to connect the two buildings, so there was no mixing of the two worlds. In fact, the only people who spent time in both buildings were the medical students, so Dr. Kabat-Zinn's introduction to the progressive-minded family medicine doctor who supplied the first cohort of meditation patients was arranged by a med student (hint: he wrote this book). UMMS was a new school (its first class had just graduated in 1974) and so its institutional identity hadn't yet had time to congeal. It was willing to give this new venture an opportunity, and the rest is history. Jon led the effort by which UMMS came to house the nation's first Division of Mindfulness in Medicine and Society, and literally thousands of papers have been written since then about the impact of meditation on health and illness.

One other forward-thinking move by UMMS was to support my interest in this area by funding my trip to San Diego in 1978 for the annual meeting of the American Holistic Health Association. I had won fourth prize(!) in an essay competition sponsored by the American Medical Student Association in conjunction with the AHHA, and going to the conference showed me that there was in fact a whole 'nother world out there (mostly in California, unfortunately), where holistic medicine wasn't quackery but a hot new field led by MDs, PhDs and other highly-trained health care professionals. It was like a peek into the future, into one possible (and maybe even probable) reality for the rest of us.

Residency and My First Job

I was able to tie these professional interests together by doing a psychiatry residency in Central California in the early 1980s. It was an interim period in psychiatry's life cycle, as medication management wasn't yet king of the hill, and there was still room for psychiatrists to perform psychotherapy (that job is now usually deferred to the psychologists). I was fortunate to train under Dr. George Solomon, the father of psychoneuroimmunology (PNI)—a jargon term describing how stress can affect our immune function, a radical notion then, but commonplace now. The only mind/body treatment I felt comfortable using with patients then was simple progressive muscle relaxation; energy medicine wasn't yet in my toolkit. I was able to get an inside view into how the mind influences the body, learn about libido and psychological defense mechanisms from my supervisors (many of whom were Freudian analysts), and even attend a seminar on hypnosis for the residents-in-training. Beyond that, it was a fairly conventional four-year program.

Next up was my first "real" job, as a staff psychiatrist at the Boston Veterans Administration Outpatient Clinic for the rest of the 1980s.

The primary diagnosis we treated was, not surprisingly, post-traumatic stress disorder (PTSD) from World War II and Vietnam, but the treatment approaches were pretty primitive in retrospect. Apart from some Valium-type meds for anxiety relief, the favored treatment approach was a form of group therapy in which the vets were encouraged to talk about their traumatic experiences. Unfortunately for them, this talk therapy didn't achieve its goal of "getting it out of their systems." Rather than defusing memories, the vets' wartime emotions were simply reactivated through reliving their experiences. So while bonding with their new squad of military buddies in the therapy group was important to these vets, their symptoms didn't improve significantly, largely because they were getting retraumatized every Tuesday morning at 10:00 a.m. I did, however, learn firsthand about the phenomenon of vicarious traumatization—the process by which a caregiver gets traumatized simply by hearing vivid descriptions of other people's traumas.

It was to be another twenty years before I learned effective ways of addressing that problem, in part thanks to the work of one of my VA colleagues, Dr. Bessel van der Kolk. He left the clinic about a year after I began and went on to write a highly influential book on PTSD based in part on his work with these patients. By calling it *The Body Keeps the Score*, he emphasized that words aren't enough to heal. But back in 1985, when he handed me the reins to his combat therapy groups as he left the VA, these ideas hadn't yet taken shape.

While mired in this therapeutic limbo, I was fortunate enough to meet my own team of like-minded buddies during those VA years, thanks to an ad I placed in the personals column of the *Psychiatric Times*. This may sound like the plot from a bad sitcom, but I was so desperate to find other psychiatrists who shared my holistic POV that I placed that ad alongside the job opportunities and help-wanted spots asking if there were other holistic psychiatrists in the Boston area who wanted to meet up. (This was how things were done in the days before social media.) I actually got four replies, and the lunchtime get-togethers we had in the ensuing years were surprisingly important in maintaining my own mental health. As was my discovery of the American Holistic Medical Association (AHMA), whose 200+ other physician members had also caught the holistic bug. Their annual meeting was a "gathering of the clan" and a big morale booster, although the psychiatrist members could be counted on one hand (Dan, Scott, Scott, and me).

Energy medicine made a cameo appearance at the VA in the form of some visiting lectures on acupuncture and a few behind-the-scenes conversations with a psychology trainee about his own acupuncture studies. But there wasn't enough staff expertise to anchor acupuncture, or any other form of EM, in the practice patterns of the very highly bureaucratized VA system. I left the VA in 1990 with lingering guilt about my inability to help my patients. Many years later (2018), I dedicated an article to them about the use of EFT for PTSD: "If only I had known then what I know now." So, it's been thrilling to read of the VA's newly created Coordinating Center for Integrative Health under the direction of a former AHMA rising star, Ben Kligler— better late than never!

THE 1990s
Getting Started at Spaulding

In 1990 I started working at Spaulding Rehabilitation Hospital (SRH) in Boston as a staff psychiatrist in the pain management program. This four-week inpatient program was started in the 1970s by a colleague of Dr. Norm Shealy, the holistic neurosurgeon and PhD psychologist who cofounded the AHMA and who pioneered the use of mind/body therapies to help manage pain. The SRH program used a team of physical therapists, occupational therapists, behavioral health practitioners, and pain doctors to help patients to maximize their everyday function. Pain relief was not the program's goal, though it was a common "side-effect." Since our chronic pain patients were not being helped very much by conventional medicines and injections, we had a green(ish) light to explore the use of mind/body therapies like meditation, hypnosis, and biofeedback (Dr. Shealy's top choice), and eventually even energy therapies like acupuncture. I was now in a favorable clinical niche and had enough professional standing as a staff physician that I could begin to introduce energy medicine concepts into our hospital. Similar trajectories were unfolding in other medical centers in America, but the developments at SRH nicely illustrate the process by which a paradigm shifts, as this decade-by-decade summary of some of my favorite bumps in that road will show.

The Celestine Prophecy—Our New Age Catalyst

This New Age bestseller about the spiritual awakening of an unemployed American school teacher (Redfield 1994) was so popular among my outpatient therapist colleagues that our lunchtime discussions of the book launched a twenty-five year tradition of after-hours meetings to share our interests in "far out" topics like that book and its Nine Insights. (True confession—I got halfway through this travel/adventure story about the discovery of ancient mystical teachings before realizing it was a work of fiction!) Even so, the book provided a compelling and accurate outline of the many dimensions of life energy and is still a useful teaching tool today. Our group members were all dually trained, having the standard professional OT, PT, RN, and MD degrees to go along with acupuncture certification, tai chi mastery, and energy healing training. So along with our Celestine enthusiasm, we had enough professional credibility to balance out the "woo-woo" factor and begin to implement these treatments. As we began to appreciate how much these new modalities helped our chronic pain patients, we formed the Integrative Medicine Task Force (IMTF) to spread the word within our hospital network of inpatient units and outpatient clinics. Our first-choice name for our group, the Spaulding Rehabilitation Hospital Integrative Medicine Project, was rejected as soon as we wrote out the acronym (SRHIMP!). Eventually we grew to have over one hundred participating staff members throughout the Spaulding system, at both inpatient and outpatient settings.

The Catch-22 of Homeopathy

SRH today has one of the top-funded research programs in the entire field of Physical Medicine and Rehabilitation (PM&R), but back in the early '90s the hospital didn't yet have a Harvard affiliation and had never received any grant money for research. Then, in 1992, Spaulding's first-ever National Institutes of Health (NIH) grant was funded by the newly formed Office of Alternative Medicine (OAM) for the token amount of $30,000. The project was designed to study the effectiveness of homeopathy in treating mild traumatic brain injury, and I served on the Institutional Review Board that ensured the study patients were fully informed about the risks and benefits of their participation. Because of the OAM's constraints on study design, the researchers were limited to choosing one of ten possible homeopathic remedies to treat each patient, out of homeopathy's pharmacopeia of several hundred relevant options. These constraints were common handcuffs in early EM research, with standard study designs unable to adapt to the individualized nature of prescribing holistic therapies. This would be like deciding whether antibiotic therapy is valid, based on giving penicillin to everyone with a fever (one size does not fit all). Nevertheless, the project demonstrated significant positive benefits, and the results were published in a mainstream medical journal (Chapman 1999).

This success put our hospital president (a cardiologist by training) in an impossible bind. On the one hand, he wanted to brag about his hospital getting its first-ever NIH grant, but on the other hand, he didn't want people to know that we were studying homeopathy, because in the medical world that was synonymous with voodoo quackery. At least the EM door had officially been cracked open for Spaulding.

Acupuncture Bridges the Gap

In retrospect, it makes sense that acupuncture was the energy modality that found the earliest acceptance at Spaulding, and in academia as a whole. In large part this was because PM&R already used an intervention called trigger-point injections that injected numbing medications into special points in the skin to relieve the localized muscle discomfort of myofascial pain. Even dry-needling was effective—trigger point injections with no anesthetic, just the needle insertion—making it an exact Western parallel to acupuncture needling. It even worked with some of the same points as in Traditional Chinese Medicine (figure 3.1c). SRH had two MD acupuncturists in the 1990s back when this was still an anomaly, and acupuncture proved to be an ideal bridge from West to East.

One physician acupuncturist, Dr. Joseph Audette, went on to develop the Harvard Medical School certification program in medical acupuncture, one that has trained hundreds of MDs over the years, but back in the early '90s, the only mention of "alternative therapies" at Harvard's annual pain medicine conference came in my lecture on alternative approaches to pain management. I was consigned to the last time slot on the final day of the five-day conference, when the audience had dwindled to less than 50% of the registrants, and their energy of attention had sunk to the level of Barely Detectable—it was the deadest audience I ever lectured to. But now twenty-five years later, the conference features a popular breakout workshop on medical acupuncture led by Dr. Bridget Chin of Spaulding, as well as a series of lectures on mind/body approaches to pain, herbal remedies, and chiropractic treatments that are highlighted with eye-catching photos in the marketing brochure—another dramatic sign of transformation deep in the heart of Harvard.

My Board of Inquiry

I soon began to use therapeutic touch (TT), the noncontact biofield treatment studied in the Rosa/*JAMA* paper, because I could readily administer it in privacy behind the curtains that separated the beds in Spaulding's semiprivate inpatient rooms. I noted a growing acceptance of TT among staff and patients, with positive outcomes and a few staff demos helping to move things forward. Then came an unexpected roadblock—the program's medical director, a bright, ambitious young neurologist on the academic fast track, subjected me to an institutional review session regarding my use of this energy medicine modality. He was not concerned about clinical efficacy or staff resistance. Rather, he said that he was concerned that I was misusing these TT sessions to take advantage of my (fully clothed) female patients. He implied that these encounters had sexual overtones, although he never explicitly made that charge or offered any concrete evidence for his concerns. (There were never any patient or staff complaints.)

So he convened a tribunal to assess what had been happening. Arrayed around the conference table, with him at one end and me at the other, were a dozen participants, including members of hospital administration, several clinical colleagues (and supporters) of mine, the medical chief of staff, but no patients or complainants. Once I was able to get past the Last Supper vibe of the setup, it became clear that nothing significant was going to occur, because nothing of substance had been alleged or had ever happened. It was a reputational hitjob, and the meeting ended with a clean slate for me and for energy medicine, though there was to be a surprise coda the following year.

Sometime after the tribunal, our clinical team brought to hospital administration our growing concerns about the program director's unprofessional behavior toward patients and staff (especially bullying and belittling language). As a result of this inquiry he was advised to seek employment elsewhere. We thought that was the end of this story, but the grand finale came the following year in the form of a series of articles in the *Boston Globe* that told how his live-in au pair had accused him of operating surveillance cameras in her bedroom and bathroom. Needless to say, things fell apart pretty quickly after that, as he lost his medical license after being convicted in a widely publicized trial. His behavior toward me was thus a classic example of projection: since his own inappropriate sexual feelings and behaviors couldn't be acknowledged as such, he displaced them onto me and my energy medicine practice (echoes of the sexual improprieties that Mesmer and Reich were each accused of). Fortunately, reality triumphed, and the use of energy medicine at

Spaulding was not prohibited. Nevertheless, I segued from using hands-on energy-based treatments to methods like EFT that were self-administered by the patient. I left the hands-on modalities to the PTs and OTs and RNs, where physical contact with patients was part of their standard of care (unlike psychiatry, the only noncontact branch of medicine).

The Grand Rounds Memo: What Not to Discuss

Spaulding's first-ever grand rounds presentation on complementary and alternative medicine (or CAM, as it was then called) and rehabilitation occurred in 1996 and was very well received by staff. Afterwards, my copresenter Dr. Audette and I got a memo from the hospital president, outlining areas he did not want us to explore any further. The main point was that he did "not want us to become leaders in nontraditional medicine." Topping the list of taboos was the use of magnets for therapeutic purposes, as well as the use of metaphors, noncontact therapies and mental telepathy. I resisted the temptation to tell him that I had a hunch he would say something like that, but the perception was still strong that our hospital's reputation would suffer if we became too closely aligned with CAM. Fortunately, as Spaulding came under new management during the subsequent wave of hospital mergers and academic affiliations, a strong enough research foundation arose that these edicts have all fallen by the wayside. Ironically, SRH now has a world-class neuromodulation lab to study the effects of external magnetic fields on brain function; we use metaphors so often that it's an endless waterfall, while empathy and compassion (the first cousins of mind-reading) have been officially designated as top clinical and institutional priorities across the Network.

Finally, Some Publicity

Our pain team's media debut came in the form of a front page above-the-fold newspaper article, complete with a full-color photo of me demonstrating the use of a novel energy-balancing device called the BioCircuit. Unfortunately, it was published not in the *New York*

Times, but in *La Presse,* the community newspaper in Montreux, Switzerland (circulation ~850). I had just given a presentation at a luxury hotel on the shores of Lake Geneva for the International Congress on Stress, showing how a simple device composed of copper plates and wires could balance electromagnetic charge in the body and create deep relaxation (figure 3.2). There were no moving parts or batteries, as the wires simply redistributed any unbalanced packets of energy or electromagnetic charge throughout the body. Its mechanism of action has never been proven, but a study using a sham circuit—one that didn't have any connecting wires but looked identical to a real one—showed that subjects felt deep relaxation only with the genuine BioCircuits and not the sham ones, thus ruling out any placebo effect (Isaacs 1991). The conference was my first and only venture

Figure 3.2—"Detecting and fighting against stress"

into the world of corporate sponsorship (my trip was funded by the BioCircuit's manufacturers), and I was pleasantly surprised to find that I did not end up feeling tainted. It was a win/win situation, and it demonstrated the international reach of energy therapies.

Some Recognition from Harvard

Boston was fortunate to be the home base for some of the pioneers of mind/body medicine. I got to know Dr. Herb Benson, of *Relaxation Response* fame (it was the best-validated and most widely used mind/body therapy at that time), when he agreed to be the faculty sponsor for a Harvard-approved course on alternative medicine in rehabilitation that our IMTF organized in 1998. This was particularly gratifying because two of the main presentations involved energy medicine—acupuncture and TT. I knew that the HMS continuing medical education (CME) committee only approved my application (and granted the continuing ed credits that would draw attendance) because of Dr. Benson's presence on the CME Committee. He was willing to go even further out on the limb for us by being a keynote speaker along with other local IM all-stars: David Eisenberg, who had just been featured in Bill Moyers' PBS Special *The Mind and Healing*, and UMMS's Jon Kabat-Zinn. Because of Dr. Benson's credibility and our use of the magic word "Harvard," the 1998 conference gave an important jump-start to the field of CAM in rehabilitation in general, with a particular emphasis on energy medicine.

The Revolutionary TV Panel that Wasn't

In 1999, I was able to convince the hospital to host a panel discussion on alternative medicine, largely because we included some big names from the world of health and media. On one side of the debate sat the President of the American Holistic Medical Association and the Editor of the journal *Alternative Therapies in Health and Medicine*. Presenting the opposing views would be Dr. Arnold "Bud" Relman, the editor of probably the most respected of all medical journals, the locally based

New England Journal of Medicine, plus a prominent internist from Brigham and Women's Hospital. The moderator was CBS News' chief medical correspondent, Boston-based Dr. Tim Johnson—the big name "get" that sealed the deal with hospital leadership. As icing on the administrative cake, we also designated the event as the first distinguished lecture to be named after our hospital chief (who had by then recovered quite nicely from his NIH/homeopathy ordeal).

The atmosphere in the packed-to-capacity conference room was highly charged, and the discussion became so lively that from my front-row seat I could see Dr. Relman's neck veins throbbing as he railed against the lack of evidence supporting CAM. The peak moment came when the AHMA's Dr. Rob Ivker plunked down a thick three-ring binder on the desk in front of Dr. Relman. Rob had come prepared, as the binder was filled with over one hundred research papers on CAM's efficacy, research whose existence Dr. Relman was evidently unaware of—he was momentarily speechless. As the drama unfolded (including his testy Q&A exchange with audience member Jon Kabat-Zinn), I grew increasingly more pleased that we had decided to videotape the event as I could easily imagine it having enough heft to become a PBS special (I have a good imagination). Only afterward did I find out that the cameraman (a maintenance guy from SRH with no prior video experience) had forgotten to take the lens cap off the camera! So my dreams of stardom were dashed, along with the chance to make a credible high-profile case for the emerging field of CAM, one that would have had a national impact. At least the audio track was preserved and the transcript was published in a prominent IM journal (Delbanco 2000).

THE 2000s
How Not to Fundraise for Integrative Medicine

Thanks to the generosity of the Boston-based Langeloth Family Foundation (and the behind-the-scenes influence of a board member who was also one of our SRH Trustees), we received enough funding during this decade to launch a golden age for IM at SRH. We could finally afford to invite guest speakers, set up research projects, train staff, and educate the community. But we never quite leveraged this early success into some of the seven-figure grants that other IM programs were beginning to get nationally, in part because Spaulding's top development priority had become fundraising for an entirely new 132-bed hospital facility being built across town. Requests for funding of any other projects were a distraction from the primary mission, so IM had to rely primarily on volunteer help from staff in order to grow. The impressive new $150 million hospital (all private rooms, green energy LEED-certified, and with stunning views of the harbor) opened in 2013, just in time to treat victims of the Boston Marathon bombing. To be fair, you must have a house before it can house any programs, but that funding focus set us up for a long dry spell.

Osher: What's in a Name?

Harvard Medical School received one of the Osher Family's early multimillion-dollar grants for IM and used it to establish the Osher Center for Integrative Medicine in 2001. At the first meet and greet for interested MDs in the Harvard community, only two MDs came—both were from Spaulding, Dr. Audette and I. The following year, the first open faculty planning meeting to launch this new division spent quite a bit of time coming up with an appropriate name. My suggested Division of Holistic Medicine was voted off the island as being too California flaky. By contrast, one of the finalists was the Division of Placebo Medicine (seriously!). Integrative Medicine won, and despite this initially rigid start, under David Eisenberg's leadership the program eventually grew and by 2020, there were several hundred doctors on board, with a membership roster of over one thousand clinicians of all types now listed on the Osher Center's interactive online network map. A subsequent director of Osher from 2012–2018, and an acupuncture researcher by training, Helene Langevin, PhD now heads the NIH's National Center for Complementary and Integrative Health—perhaps the country's top IM position.

Two Points of Light

Two key staffers helped bring energy work to the frontlines at Spaulding. Maureen Foye, RN, a graduate of the highly respected Barbara Brennan School of Healing, worked with our pain inpatients and was the healer in a pilot research study on energy balancing for chronic pain (Leskowitz, Foye, Maliszewski 2016), while Janice Wesley, PT was a multitrained physical therapist who worked her magic in our outpatient clinic. Both were sought out by staff for direct hands-on treatments, and word spread of the impact of these therapies. For the PM&R residents who rotated through Janice's outpatient clinic, these experiences planted many seeds in the minds (and biofields) of future thought leaders and generated more acceptance of EM throughout the network.

THE 2010s

Reiki and the Power of Volunteers

Thanks to a grant from the Boston-based Maynard Family Foundation, almost one hundred staff across the network were able to receive tuition-free training in Reiki, the most widely known of the hands-on energy therapies. They shared their skills at our monthly "Re-Energize!" lunchtime mini-treatment sessions for staff; the original name for those sessions, "Pit Stop," was deemed too degrading for our refined and elegant staff. The brief sessions were aimed at colleagues who were too overworked to take fifteen minutes out of their schedule for something as nonproductive as self-care. (That was the basic paradox, being too stressed out to practice stress management.) We couldn't convince administration to grant direct work productivity credits for attending the Pit Stop, but they did distribute free tickets to the local cinema. Treatment options also included foot reflexology (which featured barefoot senior physicians emitting not-quite-R-rated sighs of relief and release) and acupuncture (almost everyone remembered to wait until all their needles had been removed before they headed back to work). Of note, at least ten physicians went on to pursue Reiki training, something that would have been unheard of twenty years ago.

A volunteer Reiki program was also begun in collaboration with local community-based practitioners and modeled on the program developed at our sister institution, Brigham and Women's Hospital (Hahn 2014). The volunteers came to the hospital several evenings each week to give free treatments to patients, family, and staff. As of the start of 2020, over four thousand such treatments had been given at Spaulding, to strongly positive reviews. The tide is slowly turning.

Words of Caution from a Founding Father

A poignant follow-up occurred at the end of that decade, during an informal conversation I was having with a senior MGH psychiatrist about the role of invisible group energies in sports contests. Dr. Herb Benson coincidentally walked by the open office door, and we invited him to join us. I brought Dr. Benson up to speed on some of my research in measuring this intangible phenomenon, and he was clearly interested. However, he said that I ought to be careful about how far afield I went, lest I burn all my bridges and jeopardize my career in academia. I thanked him for his counsel, which clearly came from a place of fatherly concern and caring. It wasn't until the next day (isn't it always that way?) that the irony of the situation hit home. Dr. Benson had been a young cardiologist when he began his research into the Relaxation Response in the 1970s, and he had taken the very same kind of career risks that he was advising me not to try. Thankfully he hadn't taken his own advice, thereby enabling the medical paradigm to shift so significantly in the past forty years. My hope was that the same outcome awaited the field of energy medicine.

Parapsychology from Iraq

The most unusual lecture ever held at Spaulding occurred in 2018, when we hosted a talk by Dr. Laith Muhammad Al Azawe, the team parapsychologist of the Iraqi Olympic Committee (probably the most extraordinary job title I'd ever seen). Dr. Al Azawe described how he would utilize his psychic abilities to make pregame diagnoses of the athletes' energy status and then instruct them in energy rebalancing techniques to ensure peak performance (Al Azawe 2020). Remarkably, he was also able to make these energetic assessments at a distance, and he and I began tracking the accuracy of his energy-based predictions for upcoming sporting events around the world. Despite many logistical obstacles (internet service in Baghdad could be somewhat spotty, to put it mildly), I soon realized that he had a definite ability to suss out which players or teams had their energetic acts together. My joking suggestion that we make Las Vegas-style pregame bets backfired, as he was (rightfully) upset at the implication that he might even consider participating in such an unethical breach of scientific integrity for personal gain. Lesson learned.

THE 2020s
Nonlocal Healing

During the age of Covid, direct face-to-face contact between patients and staff was minimized, especially in the outpatient setting, so we were forced to get creative. We implemented a variant of Reiki called distant Reiki where the practitioner and the patient don't have to be physically or even temporally near each other for the benefits to be felt. So-called nonlocal healing has a surprisingly strong scientific foundation, including the best-known variant—intercessory prayer (Dossey 2014). For those who would like to experience this form on nonspecific healing, an audio recording on the Spaulding website connects listeners to healing Reiki energy—even though the recording was made months earlier and the practitioner was miles away from the subject. Although a complete scientific explanation for this modality isn't known, it clearly works.

Other adaptations that were implemented at Spaulding were part of the worldwide Covid-induced revolution in telemedicine. In addition to clinical interviews and follow-ups that could be done online, our team offered an array of online integrative health lectures, training, and treatment programs, in both individual and group formats. Unfortunately, there's still no way to do massage or acupuncture online!

Consulting for Hollywood

One of my all-time favorite synchronicities occurred in late 2022, when the furniture delivery man who commented on my wife's British accent, spoke of his education in London (at the Royal Academy of Dramatic Arts!), and then described his upcoming project shooting a film in a converted mill near our home. When he mentioned that it would be about vets recovering from PTSD by using new treatments like EFT and I told him of my background in energy medicine, veteran trauma, and academia. Our jaws dropped in tandem and then he invited me to be a psych consultant on the set. I was able to instruct the actress playing the psychotherapist (Oscar-nominee Virginia Madsen, of *Sideways* and *Prairie Home Companion*) in the correct technique for tapping (figure 3.3). The film, *Sheepdog,* due out in 2024 will hopefully help to spread the word about tapping even further.

LESSONS LEARNED

In summary, my story and the Spaulding story mirror the changes that were happening nationally with respect to integrative medicine in gen-

Figure 3.3—Virginia Madsen practicing her EFT protocol.

eral and energy medicine in particular. One tangible marker is the way that federal funding for the field as a whole rose to $152 million in 2020 from its initial $2 million seed money grant in 1992, so something has clearly shifted. Although we at Spaulding may have had a bumpier road than others, the same processes were at work at all levels. Our main challenges are summarized by these pop-culture quotes:

- "The times, they are a-changin." Bob Dylan. And the paradigms, they are a-shiftin', first in individuals, and then in institutions.
- "There's more than one way to skin a cat." Aesop. The energy medicine option has now become a go-to choice in many clinical situations.
- "If you build it, they will come." *Field of Dreams.* It's not always possible to build the full infrastructure, but once people know that the EM show is in town, they will come.
- "I did it my way." Frank Sinatra. Every hospital and medical school finds a unique path that aligns with the people and values that define them There's no single template for everyone.
- "That's just, like, your opinion, man." *The Big Lebowski.* And when that opinion is built on a solid foundation of research and clinical data, it has moved into the category of fact.

In other words, we were able to do IM our way because there's more than one way to skin a cat, and in our opinion, this was how best to build an IM program that patients would come to—and they did.

Notice to readers: No scientists were harmed in the shifting of this paradigm.

PART II

A Cartography of Energy Medicine

4
How to Feel Energy

INSULTS AND METAPHORS: PICK YOUR POISON

If the doors of perception were cleansed, every thing would appear to man as it is—Infinite.

WILLIAM BLAKE:
THE MARRIAGE OF HEAVEN AND HELL

We will soon be exploring the energy model and learning more about its theoretical and practical applications. But before we begin it's important to cleanse our own doors of perception by dismantling some crucial conceptual issues that prevent people from accessing a deeper understanding of the energetic nature of reality. These preconceived notions are linguistic devices that play on our emotional reactivity, psychological depth charges that have been implanted into our unconscious minds, where they can be easily triggered to distract and divert us from the truth. The emotional limbic system will win out over the cognitive cortex unless we stay vigilant. The two primary techniques are the use of pejorative language to describe and marginalize health care outsiders, and hiding old-paradigm images and metaphors inside our discourse to keep us stuck in outmoded ways of thinking. When those linguistic traps—the insults and metaphors—

have been identified and defused, we're in a much better position to absorb a new point of view. So here are the IEDs that the bomb squad will be deactivating.

Insults

When we think about the barriers to acceptance of unconventional approaches to health, it helps to remember that some of the most picturesque and emotionally loaded words in the English language were coined for the specific purpose of devaluing medical nonconformists (Barnhart 1995). We may chuckle at these terms today, but their impact persists.

Quack: from the Dutch *quacksalver,* a seller of salves who attracts attention with shouts (or "quacks")

Charlatan: from the Italian *ciarlare*—to chatter or babble.

Mountebank: from the Italian *monta in banco*—mounted atop a bench (to better hawk one's wares).

Swindler: from the German *Schwindel*—to be giddy, to act extravagantly.

Huckster: from the Middle English *hucc*—to haggle, and *ster,* feminine (someone who peddles aggressively and misleadingly—peddlers used to be predominantly female).

Hoax: from *hocus-pocus,* a sham-Latin magical incantation used by conjurers in the 1600s to distract and deceive their audience.

Humbug: British student slang from the 1750s, meaning a trick or jest. Made famous by Dr. John Warren at the first demonstration of ether anesthesia in Boston in 1846—"Gentlemen, this is no humbug!"

Bunk: nonsense—from Buncombe County in North Carolina, whose congressman inserted the county's name into an 1842 speech in Washington DC so his constituents back home would see it in the newspaper and stop wondering whether he was on the job or slacking off.

Crackpot: a crazy person—from the seventeenth century British slang "brain-cracked," later combining with "pot" (meaning skull). In this vein, skeptics jokingly refer to the field of psychiatry as "psychoceramics."

Bogus: a term used by the criminal underworld as far back as the 1820s to denote a machine used for making counterfeit coins.

Chicanery: the use of trickery—from the medieval French *chicane,* a form of golf derived from polo, a game played by Persian cavalry and royalty for over two tousand years. The crooked polo mallet was called a *chaugān* by Persian aristocrats.

Shyster: from the German *scheiss,* or shit; this term is usually reserved for lawyers, but it's too colorful to leave off the list!

It's a bumpy road from mockery to acceptance. The "bogus bunk hawked by these swindlers" has ranged from snake oil to Coca-Cola. The former was popular in the nineteenth century Wild West but is now known to contain omega-3 fatty acids that are powerful antioxidants. The latter was concocted as a temperance medicine by a Confederate colonel and doctor to deal with his own morphine addiction by transitioning to coca leaf extract (active ingredient: cocaine), flavored with, and caffeinated by kola nuts. The cocaine was eventually removed from Coca-Cola ingredients in 1929. Perhaps it's a sign of progress that we now use less emotionally charged terms like "pseudoscience" or "misinformation" to describe outliers like these. But regardless of pejorative intensity, the underlying message is the same—some treatments are "good," and some treatments are "bad," and it takes some conscious critical thinking to identify and reprogram these hidden biases.

Pick a Metaphor, Any Metaphor
More impactful nowadays than old-fashioned insults, though, is cultural programming to accept a limited view of science. Our worldview is seeded with imagery that shapes our ability to understand reality.

Bringing these unconscious images and metaphors to the surface can free us to grasp the bigger picture, so let's start dismantling some of the metaphors that hold us back.

There is no better place to start this discussion than with the Covid-19 pandemic, the major health issue of the 2020s. Our response to the virus came up short because, in a nutshell, we chose the wrong metaphor. The newspaper headlines tell the story: "the war against Covid," "finding new weapons to fight Covid," "losing ground to our implacable enemy," and so on. You'd think that the media's war correspondents had gotten transferred to the health desk, but in fact what happened was that a very exciting and attention-grabbing set of images has been adopted, and adapted, as the best way to talk about Covid-19. While this approach may sell a lot of newspapers and drive up the TV Nielsen ratings, it doesn't get to the root of the problem at hand.

One of the reasons we were stuck in our Covid bind for so long is because of the images and metaphors that we've chosen to help us understand the nature of health and illness in general, not just in this instance. Consider this example from *The Alphabet Versus the Goddess: The Conflict between Word and Image*, a book by sociologist Leonard Shlain (1999). Shlain describes two very different ways of working with reality—the cognitive, analytic, verbal world of the Alphabet versus the inclusive, synthetic, emotive world of the Goddess. It was his nonmedical way of talking about the two hemispheres of the brain: the logical and verbal left, the emotional and poetic right.

"The key tools of the Goddess are image, poetry, metaphor." It's as though at some point, our society was offered a choice: "Pick a metaphor, any metaphor. . . " It's an important choice, because once you pick a metaphor, your worldview is set in stone and your doors of perception become exponentially harder to cleanse. If you choose wisely, you'll be guided to insights and solutions. But if you choose poorly, you'll be unconsciously bound by the constraints of someone else's point of view, one which they wish to impose on you.

That's what metaphors do—they shape our perceptions and guide our conclusions, usually in ways we're not even aware of. Advertising (and political propaganda) is based on this process—implanting a point of view into the mind of the viewer or reader without their conscious awareness in order to gain control over people's behavior. These narratives are usually disempowering to an individual's sense of agency, as we shall see. How ironic, then, that the pioneer of this field of advertising and influencing, especially in the use of propaganda for political ends, was Edward Bernays, the nephew of the great pioneer of the unconscious mind, Sigmund Freud (Gruder 2014).

Surprisingly, no area of human activity is immune from these processes of unconscious manipulation—even the supposedly logical and rational world of science. Take Western medicine. As mentioned in the introduction, its underlying conceptual model overflows with mechanical imagery to the point where that perspective, and its inherent assumptions, are simply taken for granted. The unfolding Covid story demonstrates how, in the same way, military images can invade our mental territory. (See how easy that was?) These images are worth unpacking, to illuminate the path ahead in our journey, to understand the nature of life energy, and to help us choose wisely which conceptual metaphors to adopt (Leskowitz 1997).

To be fair, metaphors are impossible to avoid. As a famous linguist once noted (actually, it was my high school English teacher), "concretizing abstractions is the very nature of conceptualization." He made this comment when I objected to a metaphor used by Shakespeare (or Plato, or someone of that caliber) because it was inaccurate if taken literally. Instead, Mr. Fortier deftly showed me how I had missed the point— to appreciate the power of metaphors, analogies, and figures of speech, one has to turn off the right-versus-wrong analytic mind and let in the poetry. So what's the best metaphor for medicine to use, if the heart is in fact more than a pump, and a virus isn't the enemy? Let's try some on for size and see how they fit.

The Automotive Model

It's a rare doctor who hasn't resorted to car analogies to explain their patient's situation: installing a cardiac pacemaker becomes "Your battery has run down and needs to be recharged"; a healthier diet is "You just need to switch to a higher octane gas"; hip-replacement surgery translates as "Too much wear and tear is rusting out your joints"; and "Time for your five thousand mile tune-up" means it's time for your annual physical exam. There's a never-ending supply of automotive comparisons to health conditions, and they can be very helpful—as a first step. When these suggestions come from your auto mechanic, they make sense and should be pursued, but from a primary care doctor, they don't go far enough. Doctors focus on the car but not the driver, the vehicle but not the one in charge. And while an exceptional auto mechanic might add safe driving tips and maintenance pointers to help extend the life of your car, the car is basically disconnected from the driver—the exact opposite of the ideal collaborative relationship between you and your body.

Life energy is the missing link that connects each person to his body, the ingredient that turns thoughts and emotions into the chemical and nerve messengers that actually change the shape and function of the body. This sort of intimate interaction is obviously not possible between car and driver, and that's where the metaphor falls short (though some drivers do get quite emotionally attached to their cars, and some cars do seem to have a mind of their own). Ironically, the car/body comparison does have its uses when it comes to the world of metaphysics. When your car wears out (when you have a terminal illness), you step out of the vehicle (you die), the car is taken to the junkyard (your body is buried in the cemetery), while you go out and get a new vehicle in the process known as re-in-car-nation (sorry!). All kidding aside, I was grateful for this metaphor several years ago when sitting with a dying friend. She was not religiously observant and admitted to being afraid of her impending death. I shared this secular image with her, and because it was so familiar and so clearly parallel to her situation, she felt great reassurance from considering the possibility that she might "just" be stepping out of her old car, thereby getting a chance to stretch her nonphysical "legs."

To push the comparison a bit further, I recently pulled into our neighborhood gas station in a rental car, yet the attendant was able to recognize me despite the unfamiliar vehicle I was driving. She was seeing beyond the externals of make/model/year/color, and was able to see my deeper "essence" (i.e., my face). To push even further, two healer friends who had felt an unexpectedly strong sense of recognition when they first met. They decided that they were recognizing each other from a shared prior incarnation. Although their bodies looked completely different this time around, their essences were unmistakably familiar. As the Buddhists would say, the key to happiness is not to identify with, or get attached to, your body (the physical vehicle), but to connect with your true essence as pure awareness (the driver).

The Military Model

As mentioned, medicine today sees disease as the enemy, to be fought with the best weaponry that modern science can provide; doctors literally talk about their "therapeutic armamentarium" (drugs, surgery, or radiation). It's not just the war against Covid—we also have the "war on cancer," the "war on drugs," and in the wider society it's the "war on poverty" and so on. And the drugs are always "anti-" something: antidepressants, antibiotics, anticonvulsants, etc. Medicines are never "pro" anything, apart from "Pro-zac."

This sort of dramatic military language is very effective at getting people to wake up and take action, but it's built on an underlying fallacy whose contradiction eventually surfaces and needs to be addressed. That's because the battleground on which all these wars are fought is our body—the chemotherapy and radiation are explosive weapons that destroy healthy cells as well as cancerous ones. In effect, every medical "war" is a civil war that we wage against some part of our body, but we know from history that civil wars are the most damaging and the most difficult to recover from, whether military or medical.

Even the less extreme idea of suppressing symptoms rather than defeating or eliminating them is problematic. What if, instead of trying to simply make symptoms go away, we look at symptoms as feedback to be heeded, so that underlying imbalances can be corrected? That's the approach I learned in my career in pain management—the symptoms were messages, telling us how and where the mind/body system was out of balance. Typically though, the medical approach to pain is more like a fire truck that races to a burning house and screeches to a halt. The fireman jumps off the truck and runs inside to turn off the alarm, then gets back in the truck and returns to the station, leaving the house blazing. It sounds crazy, but muting the alarm instead of putting out the fire is what the field of pain management too often does with its nerve blocks and narcotic painkillers. It's certainly quieter without the alarm always blaring, but as I'll be describing, the more we listen in to symptoms rather than trying to suppress or defeat them, the more we learn, and the deeper the healing will be.

The Germ Theory

Louis Pasteur, the famous French scientist, is widely considered to be the father of microbiology and the driving force behind the germ theory of disease. But surprisingly, even before his development of novel vaccines against rabies and anthrax had gained him wide recognition in the late nineteenth century, vaccination against smallpox had been widely used by Ottoman Turks. The practice spread to England via the "Letters from the Ottoman Empire" written in 1718 by the wife of the British ambassador to Turkey (Flemming 2020). Vaccination was popularized by Edward Jenner almost one hundred years later using the milder but genetically related cowpox germ to trigger immunity (the word "vaccine" derives from the French *vache*—"cow"). As a sign of Pasteur's fame, the process of purifying milk by heating it to destroy latent bacteria is named after him (pasteurization).

This heroic view of men (unfortunately women were typically excluded from the narrative) using their intellect to defeat germs has held sway in our medical system for centuries, and by the early and mid-twentieth century, medical culture so overlapped popular culture that biographies of microbiologists could be international bestsellers. Paul de Kruif's 1926 classic *Microbe Hunters* was translated into eighteen languages and described a literal pantheon of medical gods, from the seventeenth-century Dutch inventor of the microscope (Antonie van Leeuwenhoek) to the American conqueror of yellow fever (Walter Reed) one hundred years ago. The protagonist in Sinclair Lewis's 1925 Pulitzer Prize-winning novel *Arrowsmith* was a fictionalized composite of these microbe hunters. When the book was adapted into a Hollywood movie, the hero invented a new vaccine that saved the West Indies from bubonic plague (though, somewhat surprisingly, he didn't get the girl). So we shouldn't be shocked if Brad Pitt reprises his *Saturday Night Live* role to play Tony Fauci in the Hollywood version of the Covid story.

The Terrain Theory

An important counternarrative to the germ theory of disease was developed by such scientists as Claude Bernard and Antoine Béchamp. This so-called terrain theory holds that host factors—the innate health of the person—provide a terrain that is more or less hospitable for germs. Germs can only take root where the soil has been weakened—by stress, poor nutrition, or lack of exercise—because they are in essence scavengers attracted to weakened or poorlydefended tissues. Otherwise, natural immunity will prevail and the "harvest" of health will be bountiful.

Another nineteenth century contemporary, Max von Pettenkofer, gave a dramatic exhibition of the terrain theory before a professional audience of scientists and medical doctors in which he held a glass of water teeming with cholera germs and said, "The terrain is everything; the germ is nothing." And then he drank it. He didn't get sick, in what was surely one of the most dramatic examples in medical history of "walking the talk." (Oppenheimer 2007).

Proponents of the terrain metaphor were not held out to be dashing heroes, because they weren't in the spotlight. Instead, they placed a strong emphasis on each person's self-care and personal responsibility for one's health. Many trends in holistic approaches to health follow this image—lifestyle medicine, mind/body medicine, integrative medicine—all focus on ways that we can each optimize our own health by changing behaviors and thought patterns. The contrast between the two models is most striking with respect to the Covid crisis, where billions of dollars have been spent developing new vaccines rather than emphasizing immune enhancement and risk factor mitigation. Why not both—not just the public health measures of social distancing and masks to limit exposure, but also building up the immune resilience of the general populations. Aerobic exercise, weight loss, and vitamin D supplementation are proven options, but are not part of the mainstream media discourse or of our public health policy. Not patentable and not profitable to Big Pharma, the cynics would say. And also not available to many Americans due to the larger societal issue of unequal resource distribution that blocks ready access to so many healthy lifestyle options.

The Energy Flow Theory

The energy view of health is that of energy as a flowing river that nourishes and heals the entire body—like the bloodstream, but invisible and intangible. When the river is blocked, whatever organ is downstream will suffer from lack of energy (depletion, lowered functioning), while whatever is upstream will get overrun with energy (inflammation, pain). Balance is the key, and treatments like acupuncture and tai chi work directly with the channels and rivulets to restore harmonious flow, first to the energy body and then to the physical body (Kaptchuk 1983; Cohen 1999).

Mechanical Approximations to Energy

For those left-brained folks who are truly wedded to mechanical images, two very simple and familiar images can make the energy model more relatable.

- **The gyroscope:** Kids have always been fascinated by spinning gyroscopes, especially the toy's ability to bounce back up again when given a knock. When the top begins to slow down though, it doesn't take much force to knock it over. The comparison is clear—a healthy person has a rapidly spinning gyroscope/body (i.e., a smoothly flowing energy system) that is resilient to stress, while a stressed-out person loses disease resistance and is more easily toppled because their bodily gyroscope is spinning too slowly. Energy keeps the gyroscope spinning, so we don't try to support a wobbly gyroscope by building external props—we find ways to provide energy to get it spinning again (in the case of the body this could be exercise, stress management, nutrition, etc.). Physicists call this balancing act "gyroscopic precession." A gyrocompass is a key navigational aid because it always points to true North. Biologists call the balancing act in a healthy individual homeostasis, although there is no consensus on where it comes from.

- **Iron filings:** I still remember how befuddled my middle school science teacher became when he tried to perform the classic science demonstration of how iron filings sprinkled on a sheet of white paper would align with the magnetic lines of force generated by the bar magnet under the paper. Nothing happened to the shapeless pile of iron dust until, in desperation, he gave the edge of the paper a tap. And then, like magic, the filings all jumped into their expected position, showing the familiar beautiful symmetry of the invisible lines of force in the magnetic field. And so it is, I believe, with the human body—the cells are like iron filings, and the biofield is the hidden magnetic template guiding the location, growth, and development of every one of our billions of cells. This simple but powerful image is my favorite of all the metaphors, and builds a foundation for the upcoming chapter on biofield medicine. Hopefully it can provide a conceptual "tap" to our stuck old ways of thinking, so that new patterns can emerge in our mental maps.

The Quest Itself

We'll continue with an image that demonstrates Western medicine's ineffective search for truth, an even more deeply skeptical image than the seven blind men or the drunkard looking for his key—Plato's famous Allegory of the Cave.

One of the best-known images in the canon of Western civilization is Plato's description of ignorant people responding to a glimpse of higher truth. In his allegory from *The Republic,* the narrator (Socrates) describes a group of prisoners inside a cave who are chained to seats around the campfire, but with their backs to the fire, so they're looking outward at the walls. They spend their days providing a running commentary about the changing shadow patterns thrown on the cave walls by the fire behind them. Their heads are in a fixed position, so they can't see how puppets behind them are being manipulated to make the shadows move as though alive. And since they have never been outside the cave they naturally take these shadows to be the essence of reality.

When one prisoner is told to leave the cave, he initially resists because the bright light outside is painful to his eyes, But his eyes adapt to the light, and he glimpses outer reality for the first time. He is delighted by his discovery and returns to the cave to tell his friends of the wonders of the light-filled world outside the cave, they naturally think he has gone mad. He encourages them to get up and leave, to see the outside world for themselves if they don't believe him. They are so convinced of his idiocy that they stay put and continue to analyze the flickering shadows on the wall.

Socrates goes on to discuss the political implications of the parallel situation, when the state is able to control the perspectives and values held by its citizens. He also describes a higher plane of reality—his world of Platonic Ideals—to show how his man in the cave faced the same dilemma that spiritual seekers have always confronted. Their mystic visions and transformational insights require too big a conceptual shift to be embraced by a mainstream that remains stuck in their caves, and so their teachings are rejected out of hand and they are forced to drink hemlock, be crucified, or burned at the stake. A famous quote to this effect is usually misattributed to Plato, but it is paraphrased from an essay by another Greek philosopher, Lucretius: "We can easily forgive a child who is afraid of the dark. The real tragedy of life is when men are afraid of the light."

Like Plato's prisoners, Western medicine is afraid of the light, and has tended to focus on a shadow realm of physical reality, rather than life's luminous energetic essence. But in recent years enough people have taken at least a quick peek at the world of light outside the cave that the other prisoners are getting restless, and an outright mutiny might not be too far off.

SENSING ENERGY

May the Force be with you.

YODA, THE JEDI MASTER
STAR WARS (1978)

Now that your preconceptions have been put aside and your doors of perception have been at least partially cleansed, let's go to the nub of the matter and experience energy directly, without any intervening concepts, labels, or metaphors. Perceiving energy is not crackpot. Clairvoyants and hippies may be the ones who talk the most about seeing auras and sensing the vibe, but every one of us has perceived energy directly at some point, even if we didn't realize what was happening at the time. The following common scenarios are presented with the understanding that there are also many other ways to perceive energy directly. It's a topic that goes far beyond the scope of this chapter, so resources are provided for those who wish to explore further. The first two exercises will help you detect the boundaries of the biofield (yours and someone else's), the next two will provide new ways of assessing energy experiences you've had many times but probably dismissed as insignificant, and the final one will help you start the day in a state of energy balance.

1. Sensing Your Biofield

The most common exercise in energy training programs teaches you how to use your hands to detect the boundary of your own energy field. Simply bring your arms out in front of you, palms facing each other, about two feet apart. Then gradually bring your hands closer together (without touching) and further apart, back and forth several times, to see what sensations you notice. Apart from any air movement or 98.6 degree Fahrenheit temperature sensations, people commonly notice a tingling feeling or a sense of pressure between the hands that's different from what they usually experience (closing your eyes will help you tune into this new sensation).

What are you feeling? It's the outer edge of your energy field, your biofield, your aura. As you bring your hands together, the outer boundary of each hand's energy balloon bumps up against the other one (like I did with the yoga teacher in chapter 3). An important piece of information is conveyed by the distance between your hands: you're measuring the size and strength of your own biofield. In fact, you can literally take your energetic temperature in this way, at any time and in any place, to gain some relatively objective feedback of what your inner state really is—when you're feeling "pumped up" versus when you're "deflated" (how uncannily accurate those figures of speech can be!). You'll probably have to repeat these exercises several times over several days to convince yourself that you're not imagining it. Or try this experiment: notice how your boundary changes with your emotions: first do some deep breathing cycles, or imagine being discouraged and sad, or remember something exciting that recently happened, before bringing your hands together. For example, I've noticed that my field expands significantly when I give demos of this process during workshops—I enjoy talking about and demonstrating the biofield, and my biofield simply reflects this inner state of enthusiasm.

One more etymological aside: the word "enthusiasm" breaks down into the prefix "en-" meaning "part of," and the suffix "-iasm" which makes the word into a noun (like chiasm or miasm). But the core of the word is the syllable "-thūs-," which was *theos* in the original Greek and meant "God." So enthusiasm is literally having the divine presence within you; not surprisingly, that sort of energy enlivens and expands the biofield. It's why holy figures throughout history and in all spiritual traditions have been portrayed as glowing or radiant—their biofields are bright enough to be felt and noticed by any and all in their presence. That's also why being in their presence can be a healing experience—it's a direct transmission of their high vibrational state.

As you play with these perceptions you'll notice that while the field covers your whole body, in different areas it will feel differently. For example, if you have a headache or a sore muscle, you'll actually be able to register a different sensation—some heat, more tingling, etc.—above that spot. This ability to differentiate energy sensations from different parts of the body is a skill that can be learned with practice and is a key part of all hands-on energy healing techniques (as in the Rosa/TT study).

It's also a skill that can be unlearned. Here's what I mean. Over the years, I've lectured on this topic to many different medically oriented audiences, ranging in age from high schoolers who were interested in possible medical careers, to pre-med students in college, med students,

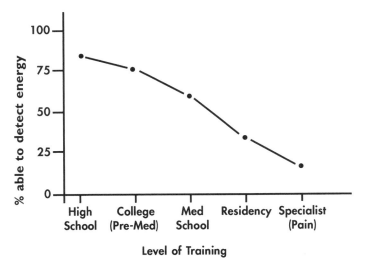

Figure 4.1—Education is a disadvantage

resident trainees, all the way up to senior physicians. I began to notice a surprising trend when I invited them to do this energy-sensing exercise. With the younger groups, large majorities of the participants could feel their energy boundaries right away. But as the participants got further along in their medical careers, the rate of successful detection dropped dramatically. Figure 4.1 illustrates the trend I saw.

These numbers are only approximate because they are based on impressionistic memories rather than actual data that I recorded at the time, (I wish I had had the foresight to jot down the actual numbers, as a more formal research study.) Nevertheless, the gist of it is quite clear: there's something about medical training that tends to "numb out" physicians and disconnect them from their bodily sensations—with one notable exception. I was pleased to discover that a group of physician acupuncturists I lectured to scored well over a 50% successful detection rate. This makes sense because they were learning to tune in to the very subtle sensations from their fingertips that would help them know when they had found the correct point for needle insertion. Being attuned in this way helped to ground them in the world of proprioception rather than cognition, an important skill for energy practitioners to develop. The decline in the MD sensitivity is not age-related, though, as energy healing workshops with non-MDs will generally have an 80-90% success rate across the age spectrum. So how did you do?

2. Sensing Someone Else's Biofield

Not surprisingly, you can also detect this field in other people. Simply direct your palms at someone else's body and do the same approach/retreat nudging or pushing against the edge of their boundaries, beginning several feet away and stopping several inches away. With their permission of course, start someplace easily accessible and emotionally neutral, like the sides of the head or shoulders rather than the face. The whole torso of a sitting person can be easily assessed, with full-body scans reserved for an actual treatment session. This assessment of someone else's biofield was at the core of the Rosa/TT study in chapter 2,

#	Chakra	Endocrine Gland	Emotion	Sensation
7	crown	pineal	bliss	scalp tingling
6	brow	pituitary	intuition	inner "light bulb"
5	throat	thyroid	creativity	"choked up"
4	heart	thymus	love	"warm-hearted"
3	solar plexus	pancreas	power	"butterflies"
2	sacral	gonads	sexuality	"turned on"
1	root	adrenal	fear	adrenaline "rush"

Figure 4.2—Sensations of chakra activation

and so the higher detection rate found by TT-trained nurses in the study prior to Rosa's (and unacknowledged in her work) makes good sense.

3. Your Energy Centers at Work

Sometimes it's hard to tell where your emotions end and your body begins. Many common emotional reactions are accompanied by strong bodily sensations that are familiar and recognizable, but which seem to be randomly located within the body. Actually, these somatic sensations have a deep inner logic that springs directly from the energy model. The knots in your stomach when you're in distress, the choking feeling when you can't express your sadness, the flash of insight that seems to come from inside your head—they each reflect the activation of one of the so-called energy centers described in many esoteric traditions, from yoga's chakras to TCM and Theosophy. But they're not all that esoteric if we allow for the possibility that powerful emotions will generate energies inside us that are strong enough for anyone, even the energy-naive, to feel. These energy charges just happen to occur in locations that are highlighted in the non-Western energy models but seem random from the anatomic/biologic perspective.

Figure 4.2 shows the key centers and the feelings we get when each center wakes up.

Let's work down the vertical axis to examine these centers in sequence:

The seventh chakra: a tingling of the scalp often accompanies an experience of awe, wonder, or unity with all.

The sixth chakra: corresponds to the "third eye" of mystic wisdom and insight, represented by internal vision or intuition.

The fifth chakra: relating to the throat—when intense emotions are hard to express, it is common to describe someone as being "choked up" as they attempt to block unpleasant truths from emerging.

The fourth chakra: the warm fuzzies often felt in the sternum are pure forms of heart energy.

The third chakra: butterflies in the stomach stem from challenges to one's personal power and autonomy.

The second chakra: being "turned on" sexually is activated in the genital region associated with this chakra, which is also related to creativity and intuition.

The first chakra: the thrill of an adrenaline "rush" is actually a whole-body activation triggered by root-chakra activation.

Subtle anatomy can explain these correspondences, but gross anatomy cannot.

4. Seeing Energy Globules

Have you ever noticed the tiny flickering curlicues of light that are sometimes visible when you look at the sky while facing away from the sun, flickers that seem to come from the clouds or blue sky itself? Not the dark circular smudges that gradually shift and sink in your field of view even when your eyes are not moving ("floaters"), and not the bright geometric patterns that appear inside your closed eyes when you rub your eyeballs (or get knocked on the head!), but the quickly moving and short-lived spiral points of light that appear against the background sky. The first two have mundane explanations: the drifting floaters are caused by dust particles or impurities floating within the eyeball itself, while the vivid geometric patterns, called phosphenes, are generated because the external pressure directly activates the retinal ganglion cells to produce visual images (Oster 1970).

What's left are those curlicues, a phenomenon that has been known for centuries and was even described by Sir Isaac Newton. Described as "vitality globules" by the Theosophists, it was most thoroughly investigated by Wilhelm Reich as part of his study of orgone energy in the skies (Reich 1951). He maintained that these flickers of light were highly concentrated manifestations of usually invisible orgone energy that were intense enough to became visible to the naked eye. He noticed certain regular patterns of globular behavior—they're more evident in good weather, in nature rather than in the city, outdoors rather than indoors. They can be seen in the dark (against your hand, for example), but most impressively, their size can be magnified by binoculars—an easy-to-do test that quickly rules out optical illusion or brain-generated hallucination as the cause. Awareness of the globules helps us realize that the sky isn't filled with empty space, but with a sort of energy that resembles the omnipresent quantum zero-point energies described earlier.

5. *Here's Lookin' at You, Kid.*

Many common idioms reflect the belief that our eyes can transmit energy: "the evil eye," "the look of love," "staring daggers at an opponent," and the sense of knowing when you're being stared at. We've all had the experience of sitting in a group setting like a restaurant or pub and turning around for no apparent reason, only to find that someone has been looking right at us. The "secret starer" isn't necessarily someone you have a magical connection with—even a random stranger can trigger this response. We typically write it off as coincidence or poetic license, but a series of controlled experiments has shown that blindfolded subjects can reliably sense when they are being stared at. Some subjects were outstanding in their ability to perceive the distant gaze, while others consciously felt nothing even as their autonomic nervous systems registered the gaze's impact via recording devices (Braud 1993).

People can develop this sensitivity with training, so in that regard energy sensing is like any other talent—there are the natural prodigies, the trainable normal folks, and the ineducable. Run your own test: have a partner sit behind you, or in front of you if you're blindfolded, and choose random times for them to look at you and then look away. You can signal when to start and finish by raising and lowering your hand. One effective demonstration at my clinic used the most skeptical member of the treatment team as the blindfolded "target" while the eight or ten of us around the table simply looked away from him when unobtrusively signaled to do so by the coordinator. The target raised or lowered his hand to show when he thought he was being stared at. It was hard to say who was more startled by the positive results, our skeptical guinea pig or the true believers.

Another option (one you might have already done during a mischievous phase of your life) is to try staring at a stranger and see what happens. Be sure to have a cover story ready, though—"I'm just doing a science experiment" might work, but then again it might not. The author takes no responsibility for any subsequent legal developments.

6. Balancing Your Own Energy

Many techniques are gaining wide popularity as methods of keeping your own energy system in balance, from the ancient art of qi gong (Cohen 1999) to modern variants like energy psychology (Greene 2009). My favorite routine borrows from several different traditions, I call it SQUEBS (Six QUick Energy BalancerS). This series of simple movements can cause a dramatic shift in your energy state in just a few minutes. To be honest, all energy techniques promise something equally impressive, so don't take my word for it or their word for it—try them out and see what happens. Measure the gap between your palms before and after the process to see how much of a change you create. Pro tip: you can enhance these effects by sighing more deeply with each exhale, even adding a vocal tone or hum to help the body vibrate and release the stuck energy (Goldman and Goldman 2017) in a sort of acoustic myofascial release.

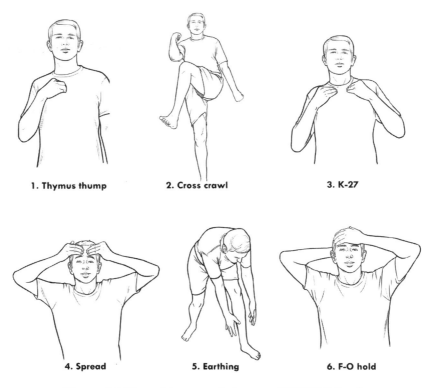

1. Thymus thump   2. Cross crawl   3. K-27

4. Spread   5. Earthing   6. F-O hold

Figure 4.3 Six quick energy balancers, SQUEBS for short.
Illustration by Rosi Fatah

1. Thymus thump—gently tapping or patting on the thymus gland (located in the center of your chest below your collarbone)—to wake up the heart of compassion (Diamond 1983).

2. Cross-crawl—to reconnect both sides of the energy body, alternating right side and left side, touch R elbow to L knee and then reverse sides, for thirty seconds, twice per second; sitting or standing (Eden and Feinstein 1998).

3. Kidney-27 Acupoint—rubbing firmly for thirty seconds on these points located underneath the center of each collarbone (Eden and Feinstein 1998).

4. Ear massage and facial smoothing—to activate key meridians, squeeze and massage entire area of both ears, including lobes. Smoothing out from center to sides, beginning at eyebrows and

working up to crown and over to back of neck and shoulders.

5. To connect to the Earth: Slowly pull energy up from the earth with palms along both sides of each leg in turn, and then distribute around abdomen (from the Dragon Tiger form of Taoist qi gong (Frantzis 2019).

6. Fronto-occipital holding (from cranial osteopathy)—Right palm against the back of the head, left palm against the forehead. With very gentle contact, allow thirty seconds and several slow breaths until release. (Upledger website 2023)

Your doors to energy perception have been opened. It's now up to you whether you cross the threshold and how far you travel on the other side. Bon voyage!

5

Understanding Energy

MORE THAN A PLACEBO: THERAPEUTIC TOUCH AND TAPPING

Now vee may perhaps to begin. Yes?

PORTNOY'S COMPLAINT,
PHILIP ROTH

The final line of Philip Roth's 1969 bestseller, *Portnoy's Complaint,* is spoken by a very patient psychoanalyst after he has listened to the main character, Alexander Portnoy, deliver a 289-page monologue describing his vast collection of neurotic grievances. Now, Dr. Spielvogel is saying (in his German/Yiddish accent), it's time to get down to work. In a way, we're at a similar point in this book. The historical context of energy medicine and the conceptual barriers that prevent its fair assessment have been outlined, so after one hundred pages we're finally ready to take a close look at some actual energy techniques and their clinical applications. We'll do so by describing the research evidence supporting two representative and well-known therapies, and then explaining how these invisible treatments work their magic on our very physical bodies. To do this we'll take a closer, and somewhat more technical, look at the anatomy and physiology of energy medicine. But first, we must do what the French Royal Commission did and eliminate any possible contributions of nonspecific factors such as the powers of suggestion and hope, so that animal magnetism (or prana or qi) is the only remaining conclusion. As Sherlock Holmes said, "Once you eliminate the impossible, whatever remains, no matter how improbable, must be the truth." (Conan Doyle 1890/1975)

As doctors today often say when nothing seems to be working and their patients are getting discouraged, "Maybe it's time for some good old-fashioned laying on of hands." Their intention is to have a heart-to-heart conversation with a supportive bedside manner that occasionally includes literal handholding. The phrase "laying on of hands," is also used by doctors as a euphemism for a practice that has been common in many branches of Christianity dating back to Jesus—the minister places his hands onto a supplicant's head and transmits the healing energy of God directly to the congregant (the Kings of England were also felt to have this power—"the royal touch"). With this figure of speech, modern doctors indirectly acknowledge that there are benefits to nonspecific factors like hope, kindness, belief and connection, although they do not admit to any healing power of prayer or priestly intervention. In other words, the phrase is a polite way of saying that all they can offer their patient, beyond medications and surgery, is the placebo effect (if you've ruled out the existence of healing energies, that's all that's left). Another etymology sidebar: *placebo* is Latin for "I will please you" and is part of the psalm known as the Vespers for the Dead, an obscure Catholic custom that evolved into the hiring of professional mourners who "sang placebos" at the bier of the deceased. By the fourteenth century the word came to mean a sycophant or flatterer. It was first used medically around 1800 to mean a harmless substance intended to "please" the patient. (Shapiro 1968), and has had a negative connotation in medicine until recent work that has detailed the powerful mind/body impact of this process (Harvard Health Publishing 2021).

Skeptics doubt the existence of healing energy in part because the treatments look so unlikely, if not outright strange—sticking needles in the skin (acupuncture), tapping on random spots on the body (energy psychology), putting hands near the body without physical contact (Therapeutic Touch), and so on. Mesmer's overly zealous use of atmospherics—his lilac silk robes, the low lighting, the eerie glass harmonica music—allowed him to optimize the nonspecific factors that we now know are such important facilitators of the healing process: hope, faith, confidence, and positive expectations. So the French Royal Society was justified in trying to weed out those psychological factors to see whether anything of substance—animal magnetism—remained. This method of analysis, known as the controlled study, has been refined over the past 250 years. We'll now look at two widely used energy-based therapies—TT and EFT—where scientific studies have definitively proven that something beyond the placebo effect is being activated.

While Therapeutic Touch (TT) involves close proximity of the therapist and patient without physical contact, the other technique (Emotional Freedom Techniques or EFT) necessitates guiding the patient in tapping on their own energy pathways. In each case, it's possible to design a treatment session that eliminates the role of personality, charisma, and expectation from the healing interaction. These research studies use a so-called "sham therapy" for comparison—a procedure that looks enough like the real thing that it would elicit a placebo response but doesn't activate the prana or qi. We'll examine some creative ways of controlling for placebo factors, a vital step in separating the truth from "whatever remains." These studies have paved the way for wider acceptance of energy medicine. Recent feature articles in mainstream media have been supportive of energy therapies like Reiki, led by *The Atlantic*'s "Reiki can't possibly work, so why does it?" (Kisner 2020) and *U.S. News and World Report*'s "Does your child need an energy healer?" (Miller 2018). So let's take a closer look at two of the therapies that have inspired this shift in public opinion.

Therapeutic Touch

Therapeutic Touch (TT) was developed through an unusual partnership between a nurse and a clairvoyant. Dora Kunz was endowed with psychic abilities since her childhood in Dutch Indonesia, seeing auras and spirits from her earliest days. She went on to become head of the largest metaphysical group in the West, the Theosophical Society of America. Kunz wanted to make her healing methods accessible to large numbers of people, even if they were not psychically gifted, because she saw it as a trainable skill. She felt this was especially important for health care workers and so teamed up with nurse educator Dolores Krieger of the NYU School of Nursing to devise a simple and straightforward protocol by which conventionally trained clinicians (especially nurses) could assess and treat energy imbalances (Krieger 1979). The protocol guided the clinician to follow a four-step process in which they will:

1. Center themselves in the present moment and enter a calm, quiet state of compassionate intent
2. Assess the client's energy field with the palms of their hands
3. Clear and mobilize the client's energy field, and then direct energy to balance the field
4. Re-evaluate the patient's field for other imbalances, treat them as needed, and then close the treatment session when energetic balance has been achieved

In a TT treatment, the nurse healer appears to touch and stroke the air several inches off the boundary of the patient's body, a process remarkably like the Mesmeric "passes" that were used to induce anesthesia two hundred years ago. If an untrained mock healer were to make the same hand movements as a trained clinician, and if subjects were unaware which sort of healer they were getting, then they would have no reason to raise or lower their expectations, in other words: no placebo effect. In many studies of TT using these sham interventions as a control, participants reliably felt a subjective sense of well-being when treated by a trained practitioner but not from the sham treatment, despite equal levels of expectation and placebo. So, as per Sherlock Holmes, the only remaining explanatory factor for any differential health effect would be the improbable one—the nurse's biofield.

TT's many clinical effects have been well documented, ranging from the psychological state of relaxation (Meehan 1998) to physiologic measures like increases in blood hemoglobin levels (Krieger 1979); to be clear though, some of these studies have been critiqued on the grounds of weak controls (Hammerschlag 2014). An estimated one hundred thousand nurses and other clinicians have been trained in TT and at least one major medical center—the University of Toronto General Hospital—has a fully-staffed TT team on call.

Sadly, the most widely publicized early study of TT—one that seemed to clearly prove that it could accelerate the process of wound healing—was later shown to be fraudulent. In this well-designed study from 1990 (Wirth), a small patch of skin was removed from the forearm of the volunteers with a punch biopsy instrument to see how quickly the wound would heal. The subjects received daily five-minute TT sessions to the affected arm, either real or sham TT, with the arm out of view behind a barrier to further minimize any expectancy and placebo factors. The lesions were photographed weekly and changes in wound surface area measured. The evaluators of the wound photos did not know whether the subjects had received real or sham TT, so their reports could not be skewed by any personal biases (this would qualify as a triple-blind protocol, but the practitioners of TT couldn't be blinded to the procedure they were administering). By day sixteen, the wounds had fully healed in thirteen of the twenty-three TT subjects, but in none of the twenty-one control subjects. The odds of this difference occurring due to chance alone were 1/1,000, qualifying the findings as statistically significant proof of a healing benefit from TT (1/20 is generally considered the threshold for statistical significance, so this effect was fifty times more powerful than chance).

Indeed, this study quickly became my favorite data point to trot out when discussing energy healing with skeptics (and believers) because it combined a solid research design with dramatically positive outcomes. At one point, I tried to contact the author, Daniel Wirth, for some additional background information, but when I never heard back from him I moved on to other projects. Several years later, a conversation with two colleagues led us to suspect that something was seriously amiss with the study. One colleague had been Wirth's thesis advisor in graduate school (that thesis was the foundation for the published article), and when he reviewed some of his old notes on his student's TT project, he concluded that the provenance of much of the original raw data had not been adequately monitored. The data, we soon concluded, had been fabricated. We published a summary of our findings (Solfvin 2005) which led to the study's deletion by the most widely respected research evaluator, the Cochrane Collaboration, from their database on noncontact healing methods (O'Mathuna 2016). Several subsequent studies by this researcher on related topics like distant healing were also retracted by his collaborators due to similar concerns about data reliability and breaches of research ethics (Carey 2004).

This was an uncomfortable project for my colleagues and me to be undertaking because the three of us were staunch believers in, and practitioners of, energy healing. We were frankly dismayed to find such clear evidence of fraud and were concerned that if our debunking was well-publicized it might deal a devastating blow to the emerging field of energy healing. Thankfully that didn't happen, because the field was already resilient enough to bounce back—just like the gyroscope in chapter 4!

Returning to the placebo effect, another way its impact can be eliminated in TT research is by using subjects who are not impacted by psychological and cognitive factors such as human infants, animals, plants, and even cell cultures. These groups all respond positively to TT and related energy therapies, and to a statistically significant degree. For example, the sprouting rate of plant seedlings was enhanced when they were soaked in TT-treated water compared to untreated water (Grad 1965), echoing Mesmer's use of baquets filled with magnetized water. In another study, infants in a neonatal intensive care unit showed a decrease of sympathetic nervous system activity following TT treatment, signifying that a state of relaxation had been attained (Whitley 2008). This is similar to the response infants show to standard soft-tissue massage therapy (Field 2010), but in TT there is no physical contact—it's a sort of phantom massage. Animals treated by holistic vets respond to a wide range of energy therapies, with acupuncture and homeopathy being the best studied (Wright 2019; Aleixo 2021), while reports of TT's use in animals is only anecdotal at this stage (NB: the word "anecdotal" in this context doesn't mean that the report is an amusing story; it's a research term for preliminary and unconfirmed case studies, the important first step to developing a larger and more rigorous clinical trial).

Several well-controlled studies done at the University of Connecticut Medical School have looked at the impact of TT on an assortment of in vitro cell cultures (i.e., isolated cells growing in a test tube filled with nutrient-rich broth). (Gronowicz 2008). In the control arm of the study, the cells were treated with sham TT, while another set of cells received true TT. Cultures of muscle, bone, and cartilage cells replicated faster when they were subjected to ten minutes of TT twice a week, compared to those cultures that received the sham TT; more frequent treatments led to more robust responses. Similar changes in DNA replication and cell differentiation have been generated by a related noncontact biofield therapy that induces changes in DNA transcription (replication) (Beseme 2018). Interestingly, externally generated magnetic fields have a similar effect on DNA, a finding that has major implications regarding the possible mechanism of action of TT, as well as the possible health effects of EMFs (electromagnetic field) from 5G cellphone signals.

Even more striking are the studies of mouse (murine) cancer and their response to an adapted form of TT. In one study, a group of mice were exposed to the dose of chemical toxin that is known to cause cancer and the death of 100% of the mice within twenty-five days. They were then given the so-called Bengston Energy Healing Method each day. The healer places his hands at the edge of a cage containing ten mice. While no direct physical contact was involved during administration of the treatment, the mice naturally gravitated toward the healer's hands at the other end of the cage. The results were striking and unmistakable—twenty-five days after their exposure to poison, not a single treated mouse had died, and none showed any signs of tumor presence (Bengston and Krinsley 2000). This intervention has not yet been studied in human cancers, but preliminary trials are now underway. The key takeaway is that no placebo factors were activated in this study, only energetics. The clinical results were clear, and in some cases quite amazing.

Emotional Freedom Techniques (EFT)

Apart from acupuncture, EFT (tapping) is the best-researched of all the modalities that fall under the energy medicine umbrella. This relatively new technique combines a self-acupressure component directly based on Traditional Chinese Medicine (the end point of each major meridian is tapped by the patient) with a classic psychological desensitization protocol (exposure to upsetting memories of gradually increasing intensity, while affirming self-acceptance) (figure 5.1).

As of early 2022, over one hundred studies have been reported in the literature, with fifty studies using some form of comparison control group to rule out nonspecific placebo factors. In addition, six meta-analyses have been published—a research format in which a series of separate but related clinical studies is evaluated as a whole for methodologic rigor and overall accuracy. This is considered the highest quality of research design, and all six showed EFT to be effective to a statistically significant degree.

In one key study of EFT, 90% of combat veterans suffering PTSD experienced a decrease of symptoms to such a degree that

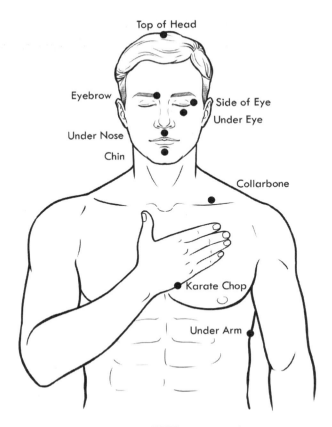

Figure 5.1—EFT tapping points
Illustration by Rosi Fatah

they no longer met diagnostic criteria after only six weeks of weekly sixty-minute EFT sessions (Church et al. 2013). The control group continued to receive standard care, but only 4% met this recovery threshold at six weeks. When the subjects in this control group crossed over and received EFT, they too showed a very high recovery rate (86%). By comparison, other widely used therapies like cognitive behavior therapy will generally report a 50% degree of symptom improvement after a longer and more intensive treatment program. Other studies of EFT show positive results in populations as diverse as high school students experiencing test anxiety and genocide survivors in Rwandan refugee camps (Feinstein 2022).

Most importantly, a recent series of studies has used a state-of-the-art brain imaging technology called functional magnetic resonance imagery (fMRI) to show that specific brain areas are reliably modulated —activated or suppressed— through EFT (Stapleton 2022). This type of study provides solid evidence of changes that match the accepted neuroscientific model of the brain mechanisms of PTSD: amygdala activation, prefrontal cortex abatement, and the like. It's an energy intervention causing a neurologic result.

Questions have naturally been raised about the mechanism of action of EFT, as so many different steps are involved in the treatment protocol. Is it the slow breathing, the muscle relaxation, the location of the points being tapped, or perhaps the affirmation of self-acceptance? In order to decide which factor is key, a series of studies must be done in which these different elements are eliminated one by one. However, it's not easy to organize and conduct these so-called dismantling studies.

For example, when I proposed a study to assess the value of EFT in treating the PTSD underlying chronic phantom limb pain, I was unable to get funding from the Harvard Medical School research initiative that I was eligible to participate in. My first proposal was a "proof of concept" design that merely attempted to ascertain whether there was any clinical benefit from EFT. However, that proposal was rejected because no plausible mechanism of action could be shown. So in my revised submission the following year, I tried to sort out the three most likely components of its mechanism of action by proposing a dismantling study. Perhaps unsurprisingly, considering the sometimes-contradictory world of academic research, this proposal was also rejected, with the reviewers citing no "proof of concept" (evidence that the treatment might actually be effective). I know a Catch-22 when I see one, so I didn't try again.

However, when EFT dismantling studies have been done elsewhere they have shown that point tapping is in fact a key component, and the specific location of the points is critical. In other words, the acupuncture mechanism is deeply involved. Similar studies with acupuncture points show that the needling of sham points does generate a mild benefit, but not to the degree seen with proper point location. So what is it about points, meridians, and all the nonplacebo aspects of EMT that we've been leading up to? Is it possible that the brain really isn't in the driver's seat? To answer this, we'll have one last look at the brain before finally diving into an updated version of the bane of every medical student's existence—anatomy and physiology.

WHAT ABOUT THE BRAIN?

I could practically see his brain cells scurrying around, forming a new thought.

B IS FOR BURGLAR, SUE GRAFTON (1985)

The heroine of Grafton's detective thriller *B is for Burglar* wasn't a neuroscientist, but her description of a suspect's thought process when caught by surprise is basically a modern take on an old philosophical chestnut—the mind/body problem. In classical philosophy, there are three main viewpoints on the relationship between the mind and the body. The *materialist* view holds that the body (in particular, the brain) generates the mind's thoughts; *parallelism* regards mind and body as complementary and interacting realms, while the *idealist* view sees consciousness as primary (like Plato's Ideal Form). In a nutshell, mainstream medicine is materialist, mind/body medicine is parallelist, and energy medicine is idealist. Let's have a quick look at each in turn, before finally exploring the energy anatomy and physiology we've been hinting at.

Materialism and Mind/Body

Over eighty years ago, neurosurgeon Wilder Penfield began a series of pioneering studies examining human behavior by applying electrical stimulation directly to the brain cortex through holes in the skull. Amazingly (and unsettlingly), this was done during awake neurosurgeries—a possibility that arises because the brain has few pain-sensing fibers of its own and so does not register pain directly; local anesthesia did the trick in these studies. He reported that stimulation of sites in the temporal lobe caused patients to recall old memories, leading him to the seemingly obvious conclusion that distinct memories were stored in specific parts of the brain and could be recalled when that spot was triggered (though how a memory could be encoded in one particular spot has never been fully outlined). It was a vivid, early example of functional neuroanatomy in action.

More recently, similar studies are being done noninvasively, as awake nonsurgical patients and study subjects receive tiny microvolt currents or milligauss magnetic pulses via scalp electrodes placed on the surface of the skin, but whose EM influence penetrates the skull and impacts the specific brain region underneath. This approach, called neuromodulation (the very technique forbidden by my hospital director twenty-five years ago) has led to many therapeutic advances in the treatment of anxiety and depression; pulsed electromagnetic stimulation (PEMS), for example, is now widely accepted in mainstream psychiatry, a far cry from the relatively primitive option of electroconvulsive shock therapy (ECT).

The neural network theory forms the conceptual basis of neuromodulation: the brain generates thoughts via interconnected and interactive networks of nerves, rather than from one particular nerve junction or synapse. One recent science bestseller embodies this model in its very title, *The Brain that Changes Itself* (Doidge 2007). The author, a neurologist, discusses the concept of *neuroplasticity*—the emerging realization that the circuitry of the brain is not permanently wired in at birth but can be profoundly modified for better or worse by life experience. Almost any set of ingrained thoughts and behaviors can, it seems, be changed, whether by trauma or psychotherapy. Thoughts can alter the brain and the brain can alter thoughts, an interactive mind/body view that has replaced the mechanistic "brain first" component of the old medical dogma.

This newer framing is eye-catching, but it falls victim to an inherent logical fallacy—an intention to change must come from a different level than the system being changed. The brain doesn't make decisions, we do—and then we use our brains to carry them out in words or actions.

Otherwise, we'd become just a "ghost in the machine" (Koestler 1967). In one dramatic example of this newly acknowledged phenomenon of neuroplasticity, regular meditation practice has been shown to make specific regions of the brain grow larger, as if the brain were a muscle that responded to certain kinds of exercise (Lazar 2005). But the "I" who decides to begin a meditation practice isn't synonymous with the "brain." "I" refers to the part of our self who has agency, and not to our neural networks, it's the self who has so far resisted all efforts to reduce it to neurons.

A helpful analogy comes from the world of computers—the owner of the computer chooses which software applications to buy and to run, and the computer just "does what it's told" as it runs the app and prints the output. But without the noncomputer element of the human being who programs the computer, nothing will happen (though AI proponents and transhumanists hope to change this balance of power). This is the dilemma faced (or avoided) by current cognitive neuroscience. Elaborate maps are made of the pathways involved in all aspects of consciousness—pain, PTSD, love, even religious experiences. (The "God Spot" is the most extreme version of this, a term describing the area in the brain that may be responsible for spiritual experiences (Bergland 2021)—but they're always couched in phrasing that never quite reveals which came first, the chicken or the egg. Here are some common neuroscience buzzwords that tend to hide this conceptual dilemma:

- Brain networks that *subserve* consciousness
- Neural mechanisms *underlying* perception
- How neural networks are *involved* in attention
- Understanding the neural *basis* of awareness
- Neurons also *regulate, modulate, organize,* and *recognize*—but where do thoughts really come from?

None of these framings truly commits to a causal path, and that's exactly where the idealist model shows its strength.

Idealism and the Mental Radio

An updated version of the world of Platonic Ideals was created by Upton Sinclair, the greatest muckraking American journalist of the 1920s and '30s. Sinclair's exposé of the Chicago stockyards in *The Jungle* spurred the federal government to adopt the Pure Food and Drug Act, he was the Democratic candidate for governor of California in 1934, and one of his many bestselling novels won the Pulitzer Prize in 1943. Sinclair was also an avid amateur student of the paranormal because his wife was psychic. He attempted to understand the nature of her telepathic abilities through a series of experiments in which he drew a picture and then asked her (across the room, facing away) to describe what he had just sketched. Their results seemed definitive enough (65 of 290 sketches were successful, 155 were partially successful, and only 70 were failures). Skeptics maintained, though, that their experimental setup wasn't airtight enough to prevent Mrs. Sinclair from picking up subtle cues from Mr. Sinclair as he sat nearby, and the criteria for a successful "hit" were somewhat subjective (Gardner 1957).

Sinclair described these tests in one of his least known works, a self-published book from 1930 called *Mental Radio*. The book justifies its title by proposing that thoughts aren't created inside the brain but are received *by* the brain, even across large distances, just like radio broadcasts. Sinclair's model aligned with the teachings of the Western metaphysical (and idealist) school called Theosophy, a philosophy that undermined the materialist view of health and consciousness by proposing that the brain is used *by* consciousness via the intervening force of subtle energy (this model will be described in more detail in the next section).

However, because Sinclair was a journalist and novelist who only dabbled in research, his views about the nature of the mind never gained traction in the scientific community. Even so, the preface to the German edition of *Mental Radio* was written by none other than Albert Einstein (Coodley 2013), a venture that embroiled Einstein in an unwarranted controversy regarding his own views on spiritualism and

ESP. During a visit to Los Angeles where he spoke of having witnessed genuine clairvoyance, the Hollywood paparazzi of the 1930s seized the opportunity for a juicy story that would sully the reputation of a great man; it took months for the controversy to die down (Grattan 1923).

An updated version of this "brain wave" model began making its way into the world of cognitive science in the early 1960s, thanks to psychiatrist Jerome Lettvin, MIT's Professor of Electrical and Bioengineering and Communications Physiology (quite the mouthful!). Lettvin's best-known research paper, "What the Frog's Eye Tells the Frog's Brain" from 1959, showed how the sensory pathways and organization of synapses in the retina structure what a frog sees by screening out types of input that just don't register (i.e., the brain is primary). But he went one step further by challenging his colleagues to take their ever-more-elaborate maps of the brain and its neural circuitry and explain where, exactly, the mental pictures of consciousness are formed. In a guest lecture at my college, I heard him compare science's current response to the challenge of explaining consciousness to the act of taking apart a TV in order to find out which component—the capacitor, the transistor, the picture tube—creates the pictures we watch on the screen. By implying that no single part of our brain creates our thoughts, he effectively presented neuroscience as "Mental TV."

One common experiment of nature seems to disprove the Mental Radio/TV model—a stroke. In a stroke, the blood supply to a particular brain region is cut off, causing permanent tissue damage due to lack of oxygen (anoxia) to that specific area, just as a heart attack damages one part of the heart muscle when its blood supply is cut off. Damage to different parts of the brain causes different types of behavioral symptoms—the best-known examples being the loss of speech when the left temporal lobe is damaged, and the loss of facial recognition when the right parietal lobe is blocked. These examples would seem to answer Lettvin's challenge—if a particular piece of equipment is removed, then certain types of TV images are lost, and certain brain functions can no longer be performed. But that doesn't mean that the function is located *inside* the nerve cluster, just as the TV picture doesn't originate inside the capacitor, or the picture tube. So where, then, does the picture come from?

Beyond Idealism

It's important to acknowledge where this train of thought leads. If consciousness isn't generated by the brain, then it isn't tethered to the brain either. A vast array of anomalous human experiences and beliefs begins to make sense within this model of radical idealism: telepathy, near-death experiences, astral travel, precognition, remote viewing, and even reincarnation. All are being rigorously studied by such scientific organizations as The Society for Scientific Exploration, the Institute of Noetic Sciences, and the Consciousness and Healing Initiative (see Resources, p. 274). Advanced graduate programs in these areas are offered by Duke University through the Rhine Institute, the University of Virginia (Ian Stevenson's Division of Perceptual Studies) and the University of Northampton (UK), which grants a PhD in Parapsychology.

This represents a huge change from my training in psychiatry, back in the days of DSM-III (the third edition of the official Diagnostic and Statistical Manual; we're now up to DSM-V). Many of these psychic phenomena were labeled as "first rank symptoms" of schizophrenia, *ipso facto* proof that someone is psychotic. The prescription was medication rather than meditation. An excellent summary of this new model of mind and science is outlined in a white paper report called the *Manifesto for a Post-Materialist Science*. It explains why we must look at the next dimension up from the brain—into the subtle world of energy and consciousness—for explanations of how the mind functions and how these energy medicine treatments work. From this expanded perspective, the brain's activity patterns will be seen to change *in response to* changes in the energetic dynamics of the whole system. They are the effect but not the cause. With that in mind (so to speak), we're now ready to outline the energetic structures that make this transition possible and will outline the complete pathway from thought to energy to physiology.

A CARTOGRAPHY OF ENERGY MEDICINE

*There are more things in heaven and earth, Horatio, than
are dreamt of in your philosophy.*

HAMLET (I.v.167–8)

The cornerstone of medical training is anatomy and physiology (A&P).
Medical students undergo their most intense emotional experience—
dissecting a human cadaver—in order to better understand the physical structures, the internal organs, that underlie health and illness. The
exploration continues at the microscopic level too, so the structure of
tissues and cells can also be clearly demonstrated. The hope, the promise, is that by knowing all these components in detail, the functioning
of the intact organism can be understood. Structure predicts function,
in other words, and by reverse-engineering an intact human body from
all the component parts, the function of the intact organism can be
best understood. Any given breakdown in function—any symptom or
disease—can be tied to an underlying broken part, or so the model goes.

But sometimes, there's no clear link medically acknowledged
between these two levels, and that's where energy finds its place as a
necessary part of the picture. Many examples prove that structure
doesn't always predict function and that standard physiology cannot
always explain observed medical events. Consider this demonstration
by early bioelectricity researcher Robert Becker. In the 1950s, he did
a series of hypnosis experiments in which a subject could be induced
to feel numbness in his forearm and hands, exactly where a long glove
would fit (Becker 1985). This is a very common demonstration that
uses guided images of cool compresses applied to the arm along with
suggestions of feeling numbness, but Becker took it to the next level
by measuring the electrical properties on the surface of the skin before
and during the trance state. He found that regions of low electrodermal
resistance exactly matched the areas of subjectively perceived numbness, with a very distinct cutoff between the two zones. The problem is

that no pattern of nerve distribution could create this boundary effect between wrist and forearm; the numbness boundary lies perpendicular to the lengthwise path of sensory nerves in the arm.

For example, when we sit too long with an arm draped over the back of a chair, the fourth and fifth fingers and the outer edge of the arm fall asleep. That matches the distribution of the ulnar nerve, so when its blood circulation is cut off by the chair, it stops relaying messages from those two fingers and the outer arm, leading to the sense of numbness. This pattern is called dermatomal, and is also seen in shingles, where the blisters crop up in bands across the upper or lower back in a band that matches the level of the spinal nerve that has been infected by the herpes virus. The (dys)function reflects the structure.

However, the skin sensations in the hypnosis demonstration do not follow a dermatomal pattern, because the nerves going along the arm do not end abruptly at the midpoint of the forearm, and there aren't any junctions or synapses along the boundary line either. There is no simple mechanism to explain how skin electrophysiology can be altered independently of neurology. Instead, we can look at it as a concrete example of the TCM aphorism: "the mind directs the qi, and the blood follows the qi." In this case, the mental image on which the subject focused established the organizational template that determined where the qi should go, or in this case, the region from which the qi should be drawn from. The blood (physiology) follows suit, when decreased qi circulation leads to impaired nerve function and the feeling of numbness.

The therapeutic interventions of energy medicine all trigger this cascade from mind to energy to body. An early non-TCM description of this link dates back to the eleventh century, when the physician Constantine the African (from Italy by way of Tunisia) described how emotions affect our body via energy: "You must avoid and reject heavy burdens and cares because excessive worrying dries out our bodies, leaches out our vital energies, fostering despair in our minds and sucking out the substance from our bones" (Lane 2023). Building on these early observations, the following examples will show in a bit more detail how various EM techniques impact different aspects of the subtle anatomic energy structures and thereby cause specific effects on the physical body. This linkage happens by a process that I have called "energy physiology" (Leskowitz 2022). By using these words, I do not mean to hijack a conventional medical term, but to suggest a deep analogy. The process by which energy fluxes affect energetic structures is strikingly parallel to the ways that physiologic changes occur in the physical body. Here then, is a closer look at subtle anatomy and energy physiology.*

Subtle Anatomy

The key elements of human subtle anatomy have already been mentioned: the biofield (aura), the energy pathways (acupuncture meridians), and the energy centers (chakras). The map that best binds them together is the Theosophical model of the seven planes of experience, which shows the hierarchy of energy frequencies ranging from dense to subtle, much like H_2O can range from ice to steam depending on its thermal energy level. So the multidimensional human being begins with the dense physical level and then rises through etheric (subtle energy), into astral/emotional, mental/causal, and finally soul and oversoul levels, with each possessing several subplanes. For example, the etheric level has three subplanes, with acupuncture energy perceived clairvoyantly as being slightly more dense than the pranic pathways (*nadis*) of yoga. Each of these frequency levels can be accessed and impacted by one of several given EM techniques.

Points and Pathways

Subjective detection and mapping of acupuncture pathways, much like my first personal experience of acupuncture (chapter 1), is not the only type of evidence for the existence of subtle anatomy. Several methods have been used to objectively demonstrate the physical aspect or imprint of the acupoints and meridians. Some of the best known are:

---

*More detailed descriptions can be found in the 2022 journal article on which this section is based, an article that contains over one hundred references for studies that are also mentioned in this chapter but are often not individually cited here.

1. Tracer dye injected into acupoints was carried by diffusion along the predicted meridian pathways in studies done in Korea almost fifty years ago (Soh 2009). They traveled in a pattern unrelated to veins and arteries, but one defined by electromagnetic gradients—not a physical structure but a functional one.

2. Electrical differences at acupoints have been measured directly by sensors, with increased electrical conductivity found at the acupoint. However, the process is not as straightforward as it sounds, because in the very act of taking a measurement, the measuring device itself (an electrode or a needle) will disturb the electrical environment. Still, a consensus has emerged that many acupoints are regions of higher electrical conductivity and decreased electrical resistance (Ahn et al. 2008).

3. Thermography to map temperature increases along meridian pathways stimulated by moxibustion (burning herbs) have produced equivocal results (Litscher 2005). Initial reports seemed to show that laser stimulation of the point above the ankle related to visual processing (according to TCM) led to increased activity in the visual cortex, even though the ankle has no nerve links to that part of the brain. Further attempts to replicate this non-anatomical connection between meridian and brain have failed, however (Kaptchuk et al. 1983).

Energy Centers

Many esoteric systems identify a series of major energy centers along the midline vertical axis of the body. The best-known system is the yoga map of the chakras, whose locations were determined during deep introspective meditation, but several more objective methods exist to identify their locations. A common method used by energy healers to assess chakra function is the pendulum test, where the swirling energy vortex of each chakra sets in motion any small object held on a string over the center (the accuracy of this method has not been reliably demonstrated). Unusual subjective sensations generated by energy fluxes often occur in places that don't match up to standard anatomical patterns but do match up to the subtle anatomy of the chakras. For example, the chart in figure 4.2 (this is on page 96) showed how certain strong emotions lead to energy sensations in places that have no clear anatomical connection to the sensation, but which do match up to the function of the associated chakra. As an illustration, why should feelings of powerlessness emerge in the pit of the stomach, rather than, say, the elbow? Or compassion in the sternum (yes, the heart is there, but why is the heart connected to love? It beats faster for many emotions, so why this specific link?). Why is sorrow described as "brokenhearted" rather than "broken brained?" And so on. . .

These chakra correspondences have been explored in detail at the subjective level by a body-oriented psychologist (Judith 2018), and at the level of magnetic field measurements by a cellular anatomist (Moga 2022). Their findings confirm that there are anomalies at the same locations that were described by the ancient yogis. There are also connections between the specific endocrine glands located in each of these areas, the hormones they release, and the emotional function of each specific chakra (Leskowitz 2022). So, several lines of evidence are converging here.

The Energy Field

The biofield is the scientific name for what laymen refer to as the aura, one's personal space. The term was constructed over thirty years ago by an NIH working group to provide a conceptual basis for the study of energy medicine, and several recent books have described this endeavor in great detail (Jain 2021; Feinstein 2024) The biofield has a definite physically measurable component—the magnetic field generated by the internal organs, especially the brain and heart. But it also has a subtle energy component comprised of concentric layers extending outward, corresponding to the planes and subplanes of the Theosophical model. These go from the densest and most easily palpable etheric layer within several inches of the skin, out through the astral layer to the mental body, a range than can extend from three to twenty feet, depending on the person and the situation.

Energy Structures and EM Modalities

Using this map of subtle anatomy, we can identify the specific energetic structures that are activated by each of the major EM modalities. Here are some examples:

EMDR: This form of psychotherapy helps patients release traumatic memories by using an odd-looking technique—the patient shifts their gaze from side to side as they follow the back-and-forth movements of the therapist's finger. EMDR is not generally considered a form of EM, and its full name—Eye Movement Desensitization and Reprocessing—reflects standard neuroscience explanations rather than any sort of subtle energy. While these explanations sound impressive, they are more post hoc (after the fact) or even wishful thinking than truly explanatory. Why, for example, should the lateral eye movements release trauma, apart from the fact that both hemispheres of the brain are involved? But from the energy POV, the eye movements are activating the sixth chakra, the center responsible for clairvoyance and psychic perceptions (that's why it's also called "the third eye"). In fact, the EMDR eye movements resemble those used in a psychic training program studied by researchers in the 1970s (Schwarz 1978) and have also been associated with increased hypnotic susceptibility. EMDR is like cartoon hypnosis, with Mickey Mouse and the swinging stopwatch. In this EMDR-induced state of sixth chakra activation, people can watch their own inner emotional process from a distance, so the emotions can discharge safely.

Heart Coherence: This HeartMath meditation focuses on the fourth chakra (the heart center) by paying attention to the sternum area while invoking emotions like appreciation and imaging that those emotions are being breathed in and out of the heart. This is a way to directly stimulate the heart-felt energies of compassion and appreciation and thereby activate and enhance the magnetic field of the heart and the biofield as a whole. As we'll see in chapter 8, this expanded biofield forms the basis for a wide range of interactive experiences in groups of people who have positive feelings for each other, phenomena such as team chemistry and fan energy in sports.

Energy healing: Many forms of energy healing—using the hands to balance energy—have been developed, ranging from classic techniques like TCM's external qi gong to Therapeutic Touch in the 1970s, and to Reiki more recently. Some modes of energy work are generalized and nonspecific, while others can target precise energetic structures. For example, Reiki has been perceived clairvoyantly to focus on the first subplane of the etheric, the most subtle of the etheric levels. Other techniques can be more precisely targeted to a wider spectrum of frequencies, with Healing Touch and Barbara Brennan Healing Science practitioners being able to detect and target specific energetic subplanes and physical organs (lungs, liver) depending on the need of the patient. Therapeutic Touch is mostly used as a general auric energizer, but can also be aimed at any specific layer of the field or a particular biologic organ. So the energy frequency and structures can be specifically targeted to match the clinical need.

Acupuncture: Acupoints and meridians are the energetic structures involved with this modality, with acupuncture needling being the most precisely aimed of all the EM interventions. The point locations were originally mapped out by Chinese healers several thousand years ago via manual palpation— seasoned practitioners, including blind acupuncturists, could sense the patterns of qi circulation with their fingertips. Acupuncture training nowadays uses anatomic landmarks to locate the points consistently, with the specific meridians and points used in any treatment session chosen in accordance with clinical need.

EFT: In Emotional Freedom Techniques (aka tapping), the subject taps a series of their own acupoints while reviewing an upsetting memory or emotion. This form of self-acupressure is designed to release the blocked qi at the emotional/etheric interface of each meridian, with all ten pathways being activated in each sequence. Additionally, before the tapping sequence ever begins, the patient states an affirmation of self-acceptance, which activates the fourth chakra energies of self-compassion and even transcendence. The statement "I accept myself fully even though I have this problem" involves two levels of self: the "I" with the problem is the personality or ego, while the "I" doing the accepting is a higher level of self, or soul. In this way, EFT acts as a bridge between the etheric, emotional, and transpersonal dimensions.

Energy Physiology

How subtle energy affects the body, or how "the blood follows the qi," is perhaps the major unanswered question in energy medicine. However, the emerging consensus is that the body's electromagnetic field (EMF)—the densest component of the biofield—acts as the intermediary between qi and biology, between the etheric and the physical bodies. The qi "steps down" in frequency to affect the EMF, and the EMF can then directly influence a wide variety of cellular functions by well-known biochemical mechanisms. Here are several areas that demonstrate this qi/EMF interface at work:

1. Oxidative Stress

Low-grade inflammation underlies many chronic illnesses and may be a direct result of impeded qi flow. Rein's microinflammation theory suggests that obstruction to qi flow in the meridians causes oxidative stress by creating free radicals that generate inflammation (Rein and Giacomoni 2010)—like rust forming within the cells (rust is oxidized iron). Antioxidant supplements help reverse this inflammatory tendency,

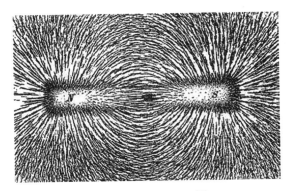

Figure 5.2a—Lines of force

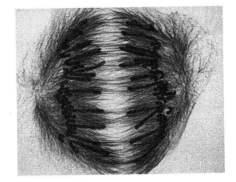

Figure 5.2b—Mitosis

as do saunas, earthing, moderate exercise, and energy medicine. Intense exercise, though, causes muscle soreness because oxidative stress (insufficient oxygen to meet the body's needs) creates lactic acid buildup and cramping. Acupuncture is known to help, as several studies of elite and student athletes have shown (Djaali 2023).

2. Organelles at the Interface

A somewhat speculative mechanism of qi-to-body interaction was suggested by the unexpected parallels between two seemingly unrelated images (figures 5.2a and b)—the familiar magnetic field around a bar magnet, and the structures inside a cell that help it to divide during mitosis, or cell division. Why do these images look so similar, even though there is no known connection between the two processes?

In this photograph of mitosis as seen under the microscope, the darker thick clumps are the chromosomes that contain the genes, while the thin strands are the mitotic spindles, a series of supportive intracellular cables that pull the structures inside the cell into their new positions. These cables are composed of microtubules, tiny structures that somehow align in the same pattern as the iron filings above the bar magnet. The problem is that there is no known discrete magnetic field, as from a bar magnet, within any given individual cell that would establish such a specific template. But according to the "inductive chain" model of energy transfer (McCurdy 2007), organelles like the mitotic spindles within the cell's nucleus (as well as the mitochondria outside the nucleus that produce cell energy as ATP) are able to utilize and transform etheric energy via this cascade.*

Each successively denser level in this chain of physical substance is induced into action by the influence of the next more subtle level: qi flow → localized EMF changes → altered cell membrane potential → subcellular organelle activation → intra-cellular ion fluxes → ATP production → biochemical activation of the cell.

Etheric energy steps down into the physical realm of magnetism via a torus, the donut-shaped vortex described by physicist Claude Swanson (Swanson 2013) as the mediator between the etheric and the physical realms; similarly, black holes pull in cosmic energy to be released into another universe on "the other side" of the vortex portal via white holes. This geometric feature could help to account for the similar shapes shared by magnetic fields and mitotic spindles, though vortices have not yet been directly visualized by electronic instrumentation. And these same intracellular magnetic forces might literally "unzip" double-stranded DNA, enabling each now-separated parallel strand to replicate and make a new set of genes.

3. Acupuncture Physiology

Acupuncture provides a clear bridge between subtle and physical energies, because the connective tissue surrounding muscles (the fascia) has physiologic properties that enable it to transmit acupuncture signaling throughout the body. The tissue planes between the layers of fascia map out many of the classic meridians (Langevin et al. 2002), and microcurrent signals along these pathways are generated by needle stimulation via the mechanism of piezoelectricity, the process that allows crystal radios to transform the physical vibrations of sound into electricity, and

---

*This summary of the way in which subtle energy affects the cell is portrayed as a cascade. The arrows each indicate a step in the process, an interaction by which the previous, more subtle, level of energy transmits its input to the next, more concrete, level of energy.

vice versa. As mentioned, electricity appears to flow more freely in these meridians by virtue of the skin's higher conductivity at the acupoints, sending pulsed electromagnetic signals to the target cells elsewhere in the body that respond to the information encoded in these patterns. Other forms of physical energy can be used to activate acupoints including light (via laser acupuncture), sound (with tuning forks on the skin), direct electrical current (electro-acupuncture), heat (burning the herb *moxa* over the acupoint—moxibustion), and even tiny magnets taped to the skin. In other words, many forms of physical energy can interact with qi to create physiologic change.

4. Disease-Specific Energy Shifts

In several variants of chronic pain syndrome, the type and location of the pain match up with the pattern of energy imbalances described by EM practitioners. As described in chapter 2, the classic acupuncture points are often, but not always, colocated with myofascial trigger points; too much qi has gathered in these spots, and acupuncture can discharge the excess energy (a process called sedation). Conversely, the tender points of fibromyalgia are treated with energizing (tonifying) needle placements because this disorder is seen as an energy deficiency. In a similar way, TCM relates specific emotions to specific meridians—grief/lungs, anger/liver, fear/kidneys, so appropriate choice of meridians will guide the healing energies to the ailing organs.

Another example of a disease-specific energy shift is seen in seasonal affective disorder (SAD), the winter "blahs." SAD is most common in the northern latitudes, because of the decreased exposure to sunshine or access only to sunshine that is low in vitality globules (see for yourself this winter). What has been called *solar prana* is the missing ingredient and can be supplied with the therapeutic light systems that are commonly prescribed. The flip side of this syndrome of winter solar energy depletion would be "spring fever," where the increasing levels of solar prana temporarily overwhelm the body's (and mind's) regulation system with bursts of energy vitality that cause emotional and behavioral excess (called "hypomania" by psychiatrists).

Some EM modalities like Advanced Integrative Therapy (AIT) focus on treating physical diseases via the chakras, guided by these classic parallels: ulcers stemming from unexpressed third chakra anger, pneumonia from blocked fourth chakra grief, thyroiditis from inability to verbalize one's truth, brain tumors from disregarding one's connection to spirit, and so on. Comprehensive descriptions of the particular energy imbalances that underlie many common diseases (cancer, arthritis, multiple sclerosis, etc.) can be found in the book *The Creation of Health* (Myss and Shealy 1988), the product of a collaboration between a neurosurgeon and a medical intuitive. EFT-based approaches also address chakra/organ disease links (Burk 2023).

Unexplained Clinical Phenomena

In order to prove that cells and organs function in patterns that are not directed by nerves, blood vessels, hormones, or genetics, counterexamples must be provided—the famous "black swans" of William James. He noted that the verified existence of only one black swan is enough to disprove the assertion that "all swans are white." Similarly, if even one energy-based disease description is shown to have more explanatory power than the "idiopathic" medical explanation, it would move the energy medicine paradigm to the front of the line.

I. Phantom Limb Pain

Chapter 6 will be entirely devoted to the clinical syndrome known as phantom limb pain (PLP), so I will only briefly note the key point here: the pain that many amputees describe as coming from their missing limb is not coming from their missing nerves, nor is it being generated by their brain as a sensory memory (the neurologic model). It is an energetic signal from their biofield, a field of subtle energy that remains intact and continues to exist in the empty space where the missing limb used to be and which can be readily detected by methods to be described in chapter 6; several other inexplicable aspects of PLP will also be described there.

2. Sighs and Yawns

An energy perspective on these common altered breath patterns emerges by paying close attention to when and where these breath changes occur. This sort of close attention to subjective sensations—microphenomenology—has been used by meditators to better understand the "sigh of relief" that often follows a moment of emotional tension discharge. The expression of the emotion releases an energetic charge, and the breath is the vehicle for this release via the deeper-than-normal sigh. Yawns are commonly used as stress-releasers during times of anxiety. They involve a vasovagal nerve response as well as the release of stagnant prana. The contagious power of yawns is usually attributed to mirror neurons, brain cells that react in parallel when we see someone performing a familiar action. But a biofield mechanism could involve our direct sensing of flow of stagnant prana that is being released within the energy fields of our yawning neighbors. A blindfold test along the lines of the HeartMath demo could easily demonstrate that the invisible energy shifts generated by nearby yawners are literally sweeping us along for the ride, apart from what we can see and hear.

3. Rapidity of Symptom Relief

The EFT literature is full of case studies of significant symptoms resolving far more rapidly (minutes to hours) than with standard treatments (days to weeks); the forty-five minute healing of one case of acrophobia (fear of heights) was videotaped and posted online as a teaching demo (Feinstein 2022). PTSD in veterans can be resolved far more quickly by EFT than the memory reconsolidation model of neural regrowth would predict.

Similarly, the rapid healing of a full thickness puncture wound through the cheek—in a matter of minutes—has been documented

(Hall 2001), while cases of cancer tumors resolving overnight have also been reported by energy healers. Standard physiologic mechanisms cannot cause changes to occur this rapidly, but energy fluxes can happen instantaneously, with the body's cells following suit automatically, like the iron filings and the bar magnet. In other words, there is more to health and illness, than is dreamt of in the philosophy of energy-less materialism. And that we have a conceptual map of where the field of energy medicine is located, it is time to follow that path and see where it can take us.

PART III

Practical Applications of the Energetic Perspective

6

The Strange Case of Phantom Limb Pain

I have heard something curious on that score, sir;
how that a dismasted man never loses the feeling of his old spar,
but it will be still pricking him at times.

THE CARPENTER TO CAPTAIN AHAB, IN *MOBY-DICK*
HERMAN MELVILLE (1851)

As we've seen in Part I, life energy has a long and deep history, high-lighted most recently by its problematic treatment at the hands of Western medicine. But as energy-based therapies have gradually gained acceptance in our health care system, the door is being opened to a wider view of what energy encompasses and what it can tell us about the nature of consciousness and of reality itself. One of the clearest examples of energy's illuminative potential can be found in the world of pain management, and in particular the clinical syndrome that often follows an amputation—phantom limb pain.

The most famous amputee in American literature, Captain Ahab, suffered from phantom limb pain. Even though his leg had been gnawed off years ago by the great white whale Moby-Dick, Ahab still felt sensations coming from "his old spar." His pain was not severe enough to be disabling, and if he'd had the advantage of some newer energy medicine approaches, he might not have felt any pain at all. That's because phantom pain provides an exceptionally clear example of the crucial role of subtle energy in health—illustrating how energy imbalances can affect the brain itself. It also leaves us one technical breakthrough away from proving once and for all that humans are energy beings first, and biological creatures second. And if that's the case, then our whole way of looking at the world must be turned inside out.

To make these points, we'll first take a look at the nature of chronic pain in general, and then focus in on the phantoms—what they are, how they can be treated, and what their energetic pattern has to say about life.

CHRONIC PAIN

Chronic pain—defined as any disabling pain that lasts more than six months despite treatment—affects over fifty million American adults and causes at least $500 billion in lost productivity and health care costs each year, numbers that have grown even larger in recent years because of the ongoing and overlapping epidemic of opioid abuse. To be blunt, America is losing its "war on pain" because the problem has been framed incorrectly. As we saw in chapter 4, if we viewed all our "enemies"—pain, cancer, poverty, terror—as symptoms instead of adversaries, we could address their underlying causes by heeding the messages they deliver, rather than just shutting off the alarm signal with pain meds and nerve blocks. Pain, and those other "enemies" provide important feedback about imbalances in our bio-psycho-social system.

But pain doctors often try to squeeze this information into a model that doesn't work anymore. For example, MRI abnormalities don't correlate with the presence or absence of pain. Only 36% of pain-free subjects in one study had normal low-back MRIs (Jensen and Brant-Zawadzki 1994), and in another, MRIs showed that over 85% of subjects without neck pain had a bulging disc (Nakashima 2015). So there was essentially no connection between pain levels and structural abnormalities—more proof that Descartes was wrong in thinking that peripheral damage causes pain, and the gate-control theory is right in seeing the brain's interpretation as key. When pain signals are treated with respect and are correctly interpreted, great change and deep healing can occur.

SELF-MANAGEMENT

Over the past thirty years, the pain management team at Spaulding Rehabilitation Hospital has developed a treatment program for chronic pain that reflects the multifactorial and nonhierarchical nature of rehabilitation medicine. Rehab is based on teamwork—every patient is seen by a physical therapist, an occupational therapist, and a behavioral health specialist, all in addition to the MD and the nurses. And at Spaulding, we've switched the focus away from reliance on medications and onto self-management. We weren't concerned with having our patients rate their pain on the common 0–10 intensity scale, which is how progress is usually measured via the recent federal guidelines—"Pain as the fifth vital sign." In fact, constantly refocusing attention on pain intensity can be a distraction, and a stressful one at that. Functional improvement was the name of our game, rather than the elimination of pain, so we tracked how many hours our patients were up off the couch, or how far they could walk without resting, or how many household chores they could do each day. We worked on such tangible goals as energy conservation, pacing, and stress management. Our multidisciplinary team acted as coaches and consultants rather than as the dispensers and source of health; we weren't the standard medical gods on a pedestal.

Chronic pain patients came to our clinic because they hadn't been able to resolve their problems with conventional medicine, where the usual approach included injections, nerve blocks, and pain killers. Our patients were desperate and willing to try anything, which is why so much important clinical work in energy medicine has been done with chronic pain patients. At Spaulding, that meant mind/body approaches like biofeedback, meditation, and hypnosis, plus energy-based therapies like acupuncture and Therapeutic Touch, were offered to all our patients, regardless of pain typology. But one form of pain proved uniquely responsive to energy approaches—phantom pain.

PHANTOM PAIN

One of the most treatment-resistant types of chronic pain—phantom limb pain (PLP)—occurs in about 30% of people who suffer amputations. These amputees are left with an intense pain that seems to be coming from their missing limb, even though they're consciously

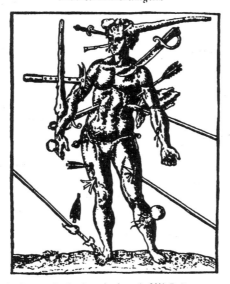

Figure 6.1—Paré's classic text

aware that the limb is missing. The pain is perceived as burning, stabbing, twisting, electrical, and searing, just as it was described by French military surgeon Ambroise Paré in his classic 1527 text on the treatment of blunderbuss injuries (figure 6.1), a cause of PLP that we never encountered at Spaulding.

Another military surgeon, Silas Weir Mitchell, was told by an American Civil War veteran: "The foot that is gone pains me the most. It seems that somebody made it their amusement playing 'stick-knife' on it a greater part of the time" (Bierle 2020).

Unfortunately, phantom pain does not respond well to medications, no matter where on the pain circuitry they're focused; opiates for the pain receptors in the skin, muscle, or bone; anticonvulsants for the synaptic relay stations in the spinal cord; anti-inflammatories for the connective tissue; tranquilizers for calming the limbic system, etc.. Fortunately, many amputees (up to 80% in some studies) do not suffer any residual pain at all, although they have the vivid sense that their limb is still present and completely intact. These "phantom sensations" are so realistic that people will forget to put on their prosthetic leg before getting out of bed in the morning. They fall down because they don't have a leg to stand on (my patient's joke, not mine!), even though it feels as if their leg was still there.

In that vein, here is Captain Ahab's response to the ship carpenter's comment quoted at the start of this chapter. The carpenter was admiring Ahab's well-crafted wooden peg leg when Ahab proposed an experiment:

> Look, put thy live leg here in the place where mine once was; so, now, here is only one distinct leg to the eye yet two to the soul. Where thou feelest tingling life; there, exactly there, there to a hair, do I. Is't a riddle? (Melville 1851, chap. 108)

In other words, the carpenter's intact leg felt just as alive to him as Ahab's phantom leg did to Ahab. Two men sensing their separate legs,

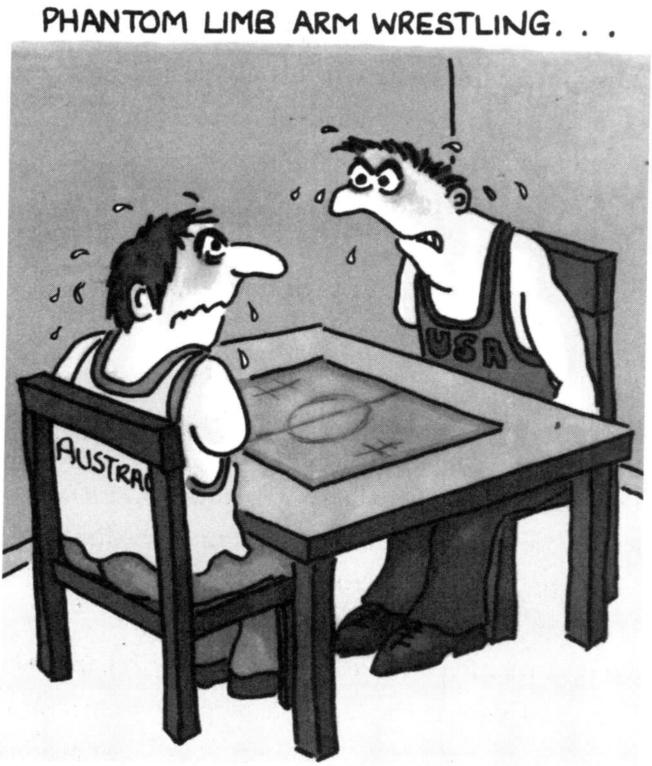

Cartoon by Terry Sedgwick

Figure 6.2—Interacting phantom limbs

both of which occupy the same location in space though only one leg is visible—a riddle indeed. The cartoon in figure 6.2 is an extreme illustration of just how real Ahab's sensation of "tingling life" is. And the case reports that follow will show that this wrestling contest is not that much of an exaggeration (though to the best of my knowledge, this sort of arm wrestling has never been tried).

It's not just limbs: phantom pain and phantom sensations have been reported following mastectomies, dental extractions, and even after the surgical removal of eyeballs and testicles. But interestingly, people who are born without a limb typically feel no pain or phantom sensation, as is also the case with children who experience amputations when still young. So, what is going on?

NEUROSCIENCE AND EMOTIONS

In the standard medical explanation of phantom pain, the brain is said to repattern and rewire itself to compensate for the nerve pathways that are no longer being used. Brain scanning with functional magnetic resonance imaging (fMRI) shows that the parts of the cortex that used to be involved in processing sensations from the missing body part have been rerouted, and now receive input from other areas of the body. This misguided repatterning process—*maladaptive neuroplasticity*—is invoked as an explanation for the anomalous sensations experienced by the patient. In other words, the brain is replaying old memories and perceptions in an attempt to maintain cognitive balance. But there is no agreed-upon explanation as to why the sensations are so painful, unless a remembered pain event had been associated with that limb (for example, if a case of gangrene led to the amputation).

The damaged nerve cells at the site of injury (the neuroma) are commonly invoked by pain specialists as the source of the pain messages. Even novelist J. K. Rowling (who would probably find some of this book's contents to be something straight out of the wizarding world), in her post-Harry Potter incarnation as writer Robert Galbraith, uses this model when she describes detective Cormoran Strike's "amputated leg stretched out in front of him, nerve endings insisting that the calf and foot were still there." (Galbraith 2020, 111). Yet if medications aimed at each stage of this sensory relay system don't work as predicted by this model, then something else must be going on. The world of the creative arts provides us with additional clues to the emotional and psychological roots of phantoms.

Novelist Stephen King's amputee protagonist in the 2008 crime thriller *Duma Key* finds—not surprisingly, considering the author—that his phantom pain triggers nightmares, psychic visions, and spirit visitations. Jodi Picoult compares the pain of emotional loss to PLP:

> I still loved him. It felt like anything else permanent that has gone missing; a lost tooth, a severed leg. You might know better, but that

doesn't keep your tongue from poking at the hole in your gum, or your phantom limb from aching (Picoult 2001).

The emotional roots of this syndrome are also addressed in a 2009 German movie aptly titled *Phantom Pain*, based on the true story of a professional cyclist whose lower leg had to be amputated after it was severely damaged when he was hit by a car. The combined stress of his resultant phantom pain and his inability to resume his cycling career sent him into a suicidal depression that was only cured by the unconditional love of his girlfriend. When his pain dissipates, that's not just Hollywood's (or Berlin's) sentimentality at work—the role of strong emotions like grief and shame in triggering phantom pain and the role of love and acceptance in healing the pain are crucial, as we'll soon see.

ENERGY MEDICINE AND PHANTOM PAIN

As mentioned earlier, chronic pain patients make ideal "guinea pigs" for novel treatment approaches: they have suffered for so long that they are willing to try almost anything. So it was when several of my amputee patients agreed to try energy medicine because nothing else was helping. Here are some descriptions of cases involving TT, Reiki, and EFT.

CASE STUDY
. . . . . .
FEELING THE PHANTOM

Jim was a thirty-five-year-old stockroom worker who had lost most of his right leg when a forklift crashed into him and crushed his leg beyond repair. He entered our program several years after his injury but did not make significant progress despite full participation in our range of treatment options. Somewhat out of desperation, I asked him if he'd like to try something completely different for his phantom pain—energy balancing. He was game, so I had him lie on his back on the exam table, close his eyes, and relax. I told him that I'd be smoothing out the magnetic field that surrounded his body, and he could either tune in or space out. I did the standard initial Therapeutic Touch assessment to trace the outer boundary of his aura and felt the same bouncy sensation off his upper body and abdomen that you might have felt in the practice exercise in chapter 5.

When, by force of habit, I continued this sensing process over the area where his right leg used to be, we both got a shock. Much to my surprise, I felt the same pressure sensation over the empty space that I'd felt over the rest of his body. And at the same moment that I was noticing this, Jim opened his eyes and said, "What are you doing?!" It turned out that, even with his eyes closed, he was able to feel my hands touching his phantom leg. The sensation was so disconcerting that he had to open his eyes to check it out. And then having done so, we resumed the session.

As I continued to smooth out his energy field with slow downward sweeps of my hands, he said that he could feel the pain draining out through a valve in the sole of his (phantom) foot, becoming less and less intense in the process. Then, out of the blue, he asked me to stop the treatment. He had realized that if the pain were to go away completely, he would have the unpleasant and undeniable awareness that his leg was truly missing. I could see that this insight would be damaging to his self-image because, as a former hockey player, his machismo had suffered as much damage from the injury as his leg had. The phantom pain was actually serving a positive function for him—it was helping him to maintain the illusion that he was in some sense intact. This psychological payoff outweighed the discomfort of his continued phantom pain, and it explained his lack of progress in our program.

In psychological terms, the pain provided him with an unconscious benefit, what's known as *primary gain*—he was able to experience a decreased sense of anxiety as a sort of emotional payoff for maintaining his ongoing physical discomfort. As the psychoanalysts would have it, his unconscious mind was doing emotional bookkeeping to balance out the physical and psychological pros and cons. He also experienced the more widely known phenomenon of *secondary gain*—he no longer had to fulfill his many roles and responsibilities at home and work, the social functions that had lost their allure for him. There was even an element of so-called *tertiary gain,* because he was now entitled to receive disability benefits through his workers compensation insurance, financial payments that he would otherwise never receive if his pain hadn't prevented him from retraining for another job. The subconscious weighing of all these psychological pros and cons led Jim to discontinue the TT; he soon left the pain program altogether and was lost to follow-up.

The sensory experience that Jim and I shared validates several key points—that the phantom can transmit sensory information to the brain, and that the phantom exists as a field or external presence in space that can be detected by others. Further, although I knew cognitively that the biofield was real, I did not expect it to be maintained in the absence of a limb, so my perceptions were not shaped by the power of self-suggestion. The psychological message of the PLP for Jim was also deep—an experience that most people would imagine to be noxious, like chronic pain, could be unconsciously transformed into something of great psychological value.

CASE STUDY

• • • • • •

PHANTOM BODY PAIN AND REIKI

Richard was a forty-five-year-old man who had suffered a spinal cord injury at the cervical level (C5–6) in a diving accident in adolescence that rendered him quadriplegic. He had enough hand strength to be able to use an electric wheelchair and was able to lead an active life, completing college, participating in adaptive sports like sailing, and pursuing a lifelong interest in Eastern religion and meditation. Recently, he had studied Reiki for self-healing and found, much to his surprise, that a generalized sense of discomfort that had always existed below his level of injury would dissipate with Reiki, leaving him with a pleasant awareness of his body's physical presence, even though he had no functioning nerves to relay these sensations to his brain. He described his pre-Reiki condition as *phantom body pain*. Neurology cannot explain his experience, since he had no functioning nerve pathways to tell his brain what was going on in his body.

Similar to Jim's experience, Richard's story also shows that energy can be perceived in the absence of nerve pathways, as can the biofield. Furthermore, energy manipulation by healing techniques like Reiki is not dependent on nerves to operate, and since these techniques cannot be fully explained by biological processes, subtle energy must be factored in.

HARVARD WEIGHS IN

Several years ago, I had the opportunity to present my work on phantom pain and energy medicine to the research team at Harvard Medical School's Osher Institute for Integrative Medicine Research. As part of the presentation, I showed them a widely viewed YouTube demonstration (D. Brown 2011) in which an amputee is able to correctly identify a series of objects placed "in" his phantom hand.

The outline of his phantom hand had been drawn on top of the table, where he'd been asked to imagine placing his phantom arm and hand. He's shocked to find that he can describe in detail the contours

Figure 6.3—Phantom hand detecting objects "in" it.

and texture of the teacup placed on top of his phantom hand's outline, even though he is looking away and has no obvious way of sensing what is being placed on the tabletop.

I then asked the Harvard group to propose possible explanations for what they had seen. I was surprised (though I probably shouldn't have been) that their comments were all highly skeptical. But it wasn't just that "he made some lucky guesses" or "the results weren't statistically significant." Instead, they wondered if "he peeked" or "he was prompted off-camera" or "the video edited out his wrong guesses." To be fair, the subject was not carefully blinded in this setup (an actual blindfold would have made for a much more convincing demonstration), but none of the researchers was willing to even entertain the possibility that sensations of subtle energies were involved in this process until every other possibility, including outright fraud, had been eliminated by the most rigorous of protocols.

The video was a true paradigm challenge because, as with Jim and

Richard, this man's phantom limb seemed to somehow be transmitting accurate sensory information to his brain. His five physical senses could not have relayed information to his brain about the objects placed on the tabletop, since there were obviously no physical sense organs in that space. To be clear, the Harvard team was comprised of leading experts in such basic science endeavors as fMRI mapping of brain function during acupuncture and analyzing the cellular micro-anatomy of acupuncture points. But the world of reductionism seeks physiologic explanations of acupuncture that don't have to invoke any sort of invisible healing energy like qi. So it was easier for them to accuse the video team of cheating than to put aside a beloved conceptual model.

Further testing along these lines would be simple enough to do. For example, Rupert Sheldrake's testing protocol from his 2002 book *Seven Experiments that Could Change the World* could be melded with Emily Rosa's TT study. An amputee could be seated behind one barrier in a row of six table-top barriers, and blindfolded energy practitioners would be asked to sense through which one of the six barriers the phantom limb was protruding. No expensive equipment would be needed, and the statistical analysis would be straightforward. As with the other six experiments he proposed, it would be simple yet paradigm-shifting. So far, though, there have been no takers.

TAPPING AWAY THE PAIN

As discussed in chapter 5, the most widely used energy-based form of psychotherapy is EFT, which can rapidly dissipate the unpleasant emotions that accompany traumatic memories, even in situations with post-traumatic stress disorder (PTSD). Given the amount of emotional distress felt by chronic pain patients, it seemed a natural approach to try. Not surprisingly, many different types of pain were helped by EFT, and its use in PLP was strikingly effective (Leskowitz 2014d).

CASE STUDY

• • • • • •

EMOTIONS AND PHANTOMS

Mary was a sixty-year-old woman who had cut her hand with a utility knife but developed such a deep infection of the skin and bones that one finger had to be amputated. The ensuing phantom finger pain made wearing her prosthetic finger unbearable, and she was so ashamed of her fingerless appearance that she became a social recluse. Using EFT, she discovered that her anger at her health insurance company for denying coverage of appropriate treatment options was even stronger and more difficult to tolerate than her pain or shame. She had spent so many fruitless hours on the phone with her claims adjuster trying to obtain coverage for pain treatment that her sense of betrayal and rage became frightening in its intensity.

However once she was able to acknowledge and accept this emotion in the therapy session by using a simplified tapping protocol, she was able to dissipate the emotional charge fairly readily. She showed up for her appointment the following week sporting a mischievous grin, saying she had something

Figure 6.4—Why is this lady giving me the finger?

she wanted to show me. I was stunned when she gave me the "F-you" hand signal, but I quickly realized that she was proudly showing off her prosthetic middle finger. It was now comfortable to wear and was accompanied by a smile of great satisfaction (figure 6.4).

Intense unresolved emotions, rather than bodily injury, sustain the energy imbalance that manifests as phantom pain. Traditional Chinese Medicine says that "the mind directs the qi, and the blood follows the qi." In Mary's case, the emotion of anger trapped the qi, and the resultant energetic pressure was so intense that it was perceivable as pain. When the negative emotion was released, the qi flowed freely again, and the pain dissipated; the nervous system followed the qi—energy physiology in action.

PHOTOGRAPHING THE PHANTOM

As intriguing as these case studies may be, scientists tend not to believe a phenomenon is real unless it can be detected and measured by a machine. Biofield researchers have developed several devices in their attempts to obtain such an image, including a camera invented over

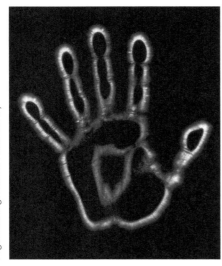

Figure 6.5—A Kirlian palm print

Nigel Hutchings/Science Photo Library

Electrode Configuration

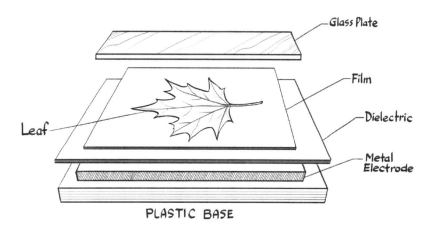

Figure 6.6—The Kirlian camera setup
Illustration by Rosi Fatah

fifty years ago by the Russian engineer Semyon Kirlian. His form of lensless photography was able to capture images of the electrical field around living objects to create eerie pictures that became part of pop culture. Figure 6.5 is similar to the Kirlian image shown in the intro segment of the hit TV show *The X-Files*.

The images themselves are generated by a form of lightning-in-miniature—the spark discharge between a charged metal plate and a living object placed parallel to its surface. Many Kirlian images over the last fifty years have shown clear evidence of an electrostatic field surrounding living organisms and inanimate objects. A leaf or coin is often used as the object to be photographed because the Kirlian equipment requires that the object should be placed as close as possible to the electric plate, as can be seen from this schematic diagram (figure 6.6). Since the discharge is not strong enough to travel more than a few millimeters, a leaf is the ideal living subject for Kirlian photography. The size of a leaf's field has been shown to increase when given infusions of healing energy and to decrease when exposed to negative emotions (Krippner and Rubin 1974).

The most dramatic series of Kirlian photos was obtained back in the 1970s by a California-based researcher who simply cut the tip off a leaf and then took a Kirlian image of that leaf. If the electromagnetic corona around the leaf was generated by the leaf tissue itself (for example, if it was created by ion flows in water within the cells), then the postpruning image should show the corona following along the truncated perimeter of the newly trimmed leaf. That is what happened in most of the images that he obtained, but in a few cases, the photographic image resembled the one shown in figure 6.7 (Krippner and Rubin 1974). The corona continued to exist in the empty space where the leaf fragment used to be!

The implications of this finding are profound, because there's no obvious, tangible source for this electrical corona tip. This time around, it's not a question of "chicken vs. egg"—the invisible EMF is independent of the physical leaf and clearly comes first, with the botanical structure somehow following suit. But how? A familiar image will help explain this mysterious finding.

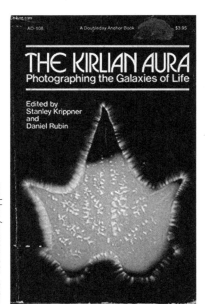

Figure 6.7—The phantom leaf effect

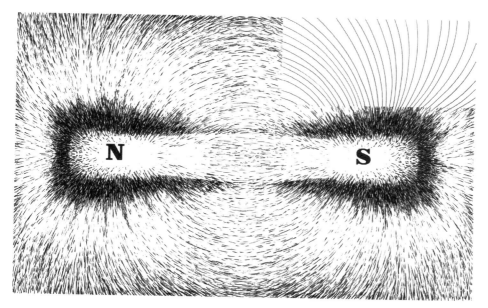

Figure 6.8—The lines of force

IRON FILINGS AND THE BIOFIELD

By now, your science teacher from chapter 5 has hopefully remembered to tap the edge of the paper, so those iron filings can snap into alignment with the invisible lines of force in the magnetic field.

It's a powerful image in and of itself, and it also provides us with a direct parallel to the phantom limb situation. Consider that if some of the iron filings are brushed away, the underlying magnetic field still exists unchanged, even though a portion of it is now invisible. In the same way, one can imagine that the cells of the human body are like the iron filings and are lined up in accordance with the invisible energy template of our subtle anatomy (the meridians, energy centers and biofield). If these cells are "brushed away" by amputation (or by leaf pruning), the underlying biofield still exists, intact but once again invisible to the naked eye.

In other words, the biofield may be the template that directs cell growth and DNA replication, the magnet that coordinates the cells' alignment behavior. This stands in contrast to an emerging view in developmental biology that cellular differentiation comes about as a result of the "self-organizing" behavior of the cells, an inherent ability of cells to use internal bioelectric signaling to direct their own specialization and patterning (Levin 2014). But again, what is the initiating cause?

As mentioned in chapter 5, studies have shown that changing the frequency and amplitude of an external magnetic field can affect many aspects of cell function, including DNA synthesis and cell function. In fact, some species that don't usually regenerate lost limbs—rats, in one study—can regrow extremities if the amputation stump is exposed to appropriate magnetic field input, in this case involving a reversal of the normal field polarity (Becker 1985). Somehow, the EMF can regulate or modulate the functions of DNA itself, perhaps by creating or entraining a resonance with the vibratory pattern of DNA's double helix and its three-dimensional conformation, as shown in one study of bioenergy healing (Rein 1995). These ideas have profound and positive implications with respect to limb regeneration (it may be possible to stimulate DNA and directly trigger limb regrowth), but negative implications with respect to the health hazards of exposure to microwave radiation; emerging evidence suggests that 5G fields may cause direct damage to DNA (Chamberlin 2022) and impair the growth of brain tissue (Singh et al. 2023).

The key implication is that if your iron filings fall out of alignment with a weakened biofield (if your cells become sick), just shifting them back to their proper location on the paper (or propping the cells up with medications) will only be a short-term solution if the field isn't also being strengthened. You, as the patient, must address any defects in the underlying magnetic field in order to permanently restore the filings (cells) to their proper alignment (healthy function). Once again, energy medicine is ideally positioned as an empowering tool that treats the underlying causes, not the externally visible symptoms.

OTHER WEIRD ASPECTS OF PLP

Validation via ESP—Well-known energy medicine practitioner and trainer Donna Eden has been clairvoyant since childhood, and as a result can see acupuncture meridians and energy fields directly. She can also see phantoms clearly, as she notes in this description of her treatment of an amputee Vietnam vet (Eden and Feinstein 1998):

I moved my hands to the ends of his legs, where his feet had been, and held the points on the bladder meridian. As the two men watched these strange conjurings, it must have seemed to them that I was just holding air. But I was not! I felt and saw the meridian lines as strongly as if his legs were still there. At first it was painful for him to have me touch the area of his absent right foot. After a couple of minutes, he reported that . . . the foot [was] being relieved of the pain.

Treating the empty space with lasers—Laser acupuncture is widely used with patients who are needle-averse, using the light from low-intensity laser pointers to stimulate the acupoints. The technique has been validated in clinical studies; for example, it is used in a standardized self-care protocol for carpal tunnel syndrome (Naeser 2006) that was developed by Boston researcher Margaret Naeser, PhD (her devices are called "Naeser's Lasers"). She and others have successfully treated phantom pain by using needles and lasers to stimulate the mirror points on the intact limb in accordance with standard TCM acupuncture protocols. More interestingly, one of her South American colleagues described his successful use of laser acupuncture to treat the phantom directly by aiming the beam at the point in space where the correct meridian points *would* be located if the limb was intact. This rather amazing result aligns with Donna Eden's experience and my own with Jim, but it was unfortunately never reported in medical literature and has not yet been replicated by other clinicians.

FUTURE DIRECTIONS OF PLP RESEARCH

Two possible lines of exploration have great potential to further strengthen the case for the biofield origins of phantom limb pain:

Laser therapy—Clinical trials of laser acupuncture for phantom pain would be relatively easy to carry out. Blindfolded amputees could be treated by direct laser stimulation of the appropriate phantom points, while inactive (sham) points, also in empty space, would be treated in control patients. Neither group would know whether they were getting sham or active treatments since both treatments look equally bizarre, so placebo and expectancy effects would be identical in both groups. Any differences in outcome could only be due to the laser's activation of the phantom acupoints.

Kirlian images—Much to the consternation of Kirlian researchers around the world, it has proven extremely difficult to replicate the original phantom leaf images obtained by Hubacher, even after fifty years of trial and error (Hubacher 2015). In part, that is because it's now impossible to reconstruct the prototype equipment, as the original components are outdated and no longer in production. Also, a certain mystique has grown around the original Kirlian operator, with some claiming that he had a unique energy field of his own that somehow facilitated the emergence of the Kirlian field around the leaf (perhaps that's what a "green thumb" actually does!).

The Oakland-based Institute for Frontier Science is currently developing a digital version of the classic analogue Kirlian camera in an attempt to reliably capture images of the phantom leaf effect, and ideally, phantom limbs themselves. Preliminary studies have helped refine the device's ability to adjust all the important parameters that will be measured (frequency, amplitude, exposure times, etc.), but no images of phantoms have yet been obtained. Another group of independent researchers is working on an updated analogue film device, with some suggestive images emerging when the researcher intentionally forms a positive emotional bond with the plant. Ideally, Kirlian cameras would be able to capture images of complex structures like phantom fingers (figure 6.9), shapes that could not simply arise as a general aura or glow being emitted from the body. Clear Kirlian images of these specific complex and biologically significant shapes would, in my opinion, provide the single most definitive and paradigm-busting proof of the biofield-as-template model.

WHY IT MATTERS

The key point of this chapter has been the demonstration, via the example of phantom pain, that the biofield is a tangibly objective structure

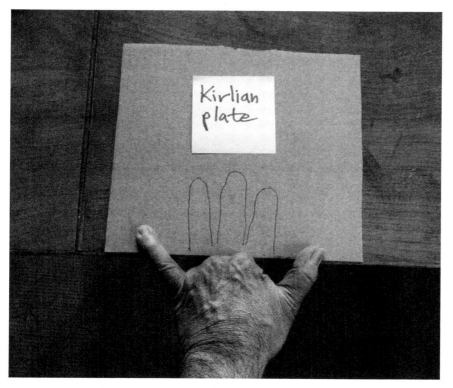

Figure 6.9—A mock-up of the ideal Kirlian outcome

that is the source of the human body's actual design and the creator of a healthy organism. The remainder of this book will describe the many implications of this fact. To that end, the next few chapters will show what happens when two or more of these biofields interact. Hint: the combination creates everything from loving couples and team chemistry to Woodstock and mass hysteria.

7
The Zone

Sport is the yoga of the West.

IN THE ZONE: TRANSCENDENT
EXPERIENCE IN SPORTS
MICHAEL MURPHY (1978)

That's certainly an attention-grabbing quote to start a chapter with. Surely it's only a hook, a nonsense phrase—football and yoga have nothing in common, right? But when you consider the source—Murphy is a polymath, author, athlete, cofounder of the Esalen Institute, and unofficial father of the Human Potential Movement (HPM)—it becomes more like a Zen *koan*, a paradox encoding deep wisdom. In fact that's what he means: the practice of both disciplines—sport and yoga—use body, mind, energy, and spirit in such integrated and coordinated ways that both can lead to spiritual awakening. There are many terms for this oft-experienced sport-related sense of energetic unity—the flow state, the Zone—that will be the subject of this chapter. We'll start with a case study about one such Zone experience.

CASE STUDY
• • • • • •

SIBLING RIVALRY AND THE ZONE

When I was eleven years old, my father cleverly exploited my sibling rivalry with my nine-year-old brother by bringing us both to the quarter-mile track at the town park and setting us up for a race. I was bigger and faster, so I wasn't really worried about the outcome, but this race was different. At the starting line, Dad pointed me in one direction and my brother in the other. It took me a minute to figure out what was going on, but as soon as the race started, I realized that I was trapped. There would be no way for me to see who was ahead until we came down the home stretch to face each other, and so I wouldn't be able to respond appropriately until the end of the race. Since the last thing I ever wanted to experience was the humiliation of losing a race against my kid brother, I decided then and there to just pour it on for the whole race, rather than save up for a final spurt.

About halfway through, something odd happened, I lost all sense of my feet and legs hitting the ground. In fact, I couldn't feel anything at all from my body because it had become just a smooth, effortless, whirling machine that was carrying me around the track. It was a pleasant floating sensation, accompanied by a sense of "Wow!" and an awareness that all my inner worries had dissolved. I totally lost track of time, but as luck would have it, I won the race by a good twenty yards and was able to hold my head high for at least a few more years. My brother soon surpassed me in all things athletic, but at least I'd had that one unforgettable experience of the runner's high—the altered state of consciousness (ASC) known by athletes everywhere as the Zone.

I never became a track star, or even a weekend jogger, but I continued to be fascinated by ASCs like that one. As millions of amateur and professional athletes have long known, it's such a compelling experience that a lifetime can be willingly devoted to its pursuit. In fact, the Zone is one of the most commonly experienced of all subtle energy phenomena, because it's not just athletes who feel it. This state is also experienced by musicians, especially improvisational ones (in the groove, cookin', in the pocket), poets (taking dictation from their Muse), martial artists (attaining *mushin*), video gamers (TV trance), poker players (in a dead stroke), and so on. However, the energy aspect of the Zone is rarely acknowledged, so in this chapter we'll look at the subtle energetics of this ASC in its physiological, psychological, and spiritual dimensions, with a special focus on sports.

Any discussion of the Zone and sports must start with Michael Murphy's 1978 classic *In the Zone: Transcendent Experience in Sports*, a compendium of Zone experiences by well-known athletes that is put into a much larger context by Murphy's deep knowledge of Eastern mystical traditions. These ASCs described by athletes are categorized by Murphy as examples of specific types of *siddhis*, the yogic term for expanded capacities and supernormal functioning. He lists twenty-one in all, ranging from telepathy and levitation (figure 7.1) to mastery of pain and transmission of energy. All of Murphy's siddhis have in common the harnessing of life force—prana—to improve mind/body coordination. He gives examples from the full sports spectrum—skiing, karate, diving, track and field, football, baseball, and more.

Figure 7.1—Levitation?

He quotes such well-known athletes as Babe Ruth, Muhammad Ali, Michael Jordan, Roger Bannister, Bruce Lee, and Jack Nicklaus as they describe experiences that can't be encapsulated in box scores or statistical analyses (and that aren't usually included in their official bios either).

Murphy is not the only researcher in this field. His original work was backed by 535 citations, and its 1995 revision grew to include 1,545 references (especially impressive, coming in the pre-Google era). Another pioneer investigator, psychologist Mihaly Csikszentmihalyi, coined the term *flow state* after hearing artists describe their total immersion in the creative process as the feeling of being carried along by a river (Csikszentmihalyi, 2014). The artists were referring to the feeling of effortlessness that is the essence of creative experience and may not even have known about qi, whose smooth flow is the key component of this state.

Because the Zone is a multidimensional place encompassing body, mind, and spirit, it can be triggered by energetic input at each of those levels in much the same way that the energy medicine modalities in chapter 6 could target specific levels of our subtle anatomy. A closer look at the inspirational and motivational potential of these dimensions will help us better understand the various aspects of subtle energy and its impact on athletic performance, as well as on creativity and even the mundane activities of everyday life.

THE PHYSICAL DIMENSION*

A mystic seeks to experience God directly, but a scientist has to be satisfied with measuring how that experience affects the physical body. So it is with the study of the Zone. As the physical is the most immediately accessible and quantifiable of the dimensions, it makes

---

*Author's disclaimer: Many of the examples I've chosen of physical interactions with the Zone focus on Boston-based athletes, since these are the folks I've followed most closely. But every city and every sport has their own pantheon of Zoners.

sense that this is where a good deal of documentation of subtle energy occurs. Many measurements of the Zone's tangible impact have been taken (especially concerning athletics) and important psychophysical correlations have been noted, with the scientific literature ranging from brief overviews (Leskowitz 2017) to in-depth review articles (Peifer 2021). Some notable areas of scientific focus and data collection on this topic are:

- **Changes in brain activity** (unsurprising, since a change in attentional focus is a key component of the Zone)—The two most important methods to measure these changes have been brain waves and MRIs. As mentioned earlier, the MRI is like a refined X-ray that can measure function as well as outline structures by showing where brain activity is highest (where blood flow is greatest). It's the gold standard for brain science, but unfortunately for the study of the Zone, the patient must lie quietly inside a big machine for the fMRI image to be taken. So we'll never get fMRIs of athletes in midcompetition, but we can at least see how the regular practice of Zone-enhancing techniques affect the brain. At least one nonathletic Zone state of high creativity—rapping—has been shown to involve a specific pattern of brain activation (Liu et al. 2012, 834)

  Meditation is the best-studied such mind/body technique, and a large body of research shows clear correlations with such important aspects of being in the Zone as increased attention, decreased distraction, and even a growth of the brain regions involved (Lazar et al. 2005), as though the brain is a muscle and the sub-units grow if they're used more. Many athletic teams now include mindfulness meditation as a standard part of their sports psychology offerings, and at least one former professional basketball player offers workshops and trainings in mindfulness under the rubric "Secrets of Pure Performance" (Mumford 2015).

Figure 7.2—NFL Coach Pete Carroll checks his qEEG readout.

- **Brain waves**—The standard tool for measuring brain waves, the electroencephalogram (EEG), is better suited to measuring meditators than marathoners. However, a portable modification contained in a headband can generate a simplified quantitative EEG analysis (qEEG) in moving subjects via a process called brain mapping. Popular models, including the Muse headband (figure 7.2) have shown that athletic performance (in one case, professional women soccer players) is predicted by qEEG testing and enhanced by qEEG training (Tharawadeepimuk 2017).

- **Changes in blood chemistry**—These are to be expected, as the hormonal system must move out of fight-or-flight survival mode to get into the Zone. For example, the release of endorphins—

the body's naturally-produced opiates (*endogenous morphine*)—had long been thought to create the runner's high of the sort that I experienced in the earlier case study. However, new studies (Bergland 2021) indicate that runners still get that high when they are given endorphin blockers (naloxone). Interestingly, a more important factor in their subjective experience was the rise of their blood levels of endogenous cannabinoids (yes, the brain has built-in receptors that recognize the key component of marijuana, a relatively recent discovery that is, for lack of a better word, far out).

- **Autonomic nervous system balance**—The impact of the flow state on our physiology is reflected by changes in heart rate variability, an index of brain/heart coordination and a key aspect of mind/body coordination. Enhanced reflexes accompany this balance, and the resulting improvement in reaction time following a heart coherence meditation allows batters to perceive an incoming fastball as though it was travelling 10 mph slower (McCraty, Tiller, and Atkinson 1996), increasing the odds that they will be able to break a slump and start a new hitting streak.

These scientific measurements are further supported through the observational and anecdotal evidence from the players themselves. Where MRIs and other brain mapping require a motionless subject, the observations below contribute a more dynamic perspective.

- **Mastery of pain**—Snowboarder Danny Kay describes how he completed his final run in the X Games mega ramp half-pipe competition despite having broken an ankle on his previous jump: "I've gotten really good at pulling the veil down, at camouflaging reality, locking out my conscious mind and riding my focus into the Zone." (Kotler 2014, 14). Even the pain of a broken bone could be "locked out" with a strong enough focus in the Zone. And while "playing through pain" can damage the body by overriding warning signals of injury, it can also lead to states of transcendence.

- **Mind/body coordination**—The physical coordination attained in these flow states can be a thing of beauty to watch. Figure 7.3 shows Red Sox outfielder Andrew Benintendi catching a fly ball during the 2018 baseball World Series.

Not surprisingly, this image was voted one of the best sports photos of the year by *Sports Illustrated* for exemplifying the grace and balance that can emerge in sporting competitions (even in events that aren't usually considered particularly elegant). Benintendi embodied what famed ballerina Isadora Duncan said one hundred years earlier: "The body, by force of the soul, can in fact be converted to a luminous fluid" (Murphy 1971). "Benny" was in a literal flow state at the moment pictured.

Jim Davis/Boston Globe/Getty Images

Figure 7.3—Baseball as ballet

THE SPIRITUAL DIMENSION

The experience of oneness beyond the self lies at the heart of the Zone experience, and is eloquently described in Eugen Herrigel's 1948 classic *Zen in the Art of Archery*.

> The archer ceases to be conscious of himself as the one who is engaged in hitting the bull's-eye which confronts him. This state of unconscious is realized only when, completely empty and rid of the self, he becomes one with the perfecting of his technical skill.

This is a more philosophical description of the commonly reported experience that mindful athletes report of no longer being distracted by critical thoughts, of being totally immersed in the event itself. Selflessness, transcendence of self, is the spiritual term for that experience.

True extra-sensory perception, gathering information not available to the five physical senses, often emerges in these transcendent states. Several forms have been described.

Precognition and Clairvoyance

Because the Zone experience is one of disengagement of attention from our normal thought processes, our mental radio becomes able to tune into other stations, including paranormal ones. During an interview about team chemistry, Boston Red Sox pitcher Mike Timlin told me that he occasionally entered a state of mind in which:

> You have the ability to feel what other people are feeling. When you're in the Zone, so to speak, out in the field you can actually tell what the hitter wants to do and doesn't want to do. We all have that capability, we all have that sense. People call it ESP, or whatever they call it, but you have that extra sense sometimes where it's a feeling in the gut or the voice in the back of your head telling you what's going on (*The Joy of Sox* 2013, 09:15).

Hockey great Wayne Gretzky put it more succinctly: "'I just go to where the puck is going to be.' For him it was not just a matter of calculating angles, but of literally seeing what will happen to him in the future." This quote is from a scientific article with the rather amazing title of "Precognition in Elite Sports" (Pates 2021).

NFL linebacker David Meggyesy once found himself playing a game

in a kind of trance where I could sense the movements of the running backs a split second before they happened. With this heightened sense of anticipation, I played a brilliant game. The speed of the game slowed down and I had a sure "knowing" where the next play was going. I saw vivid colors and patterns of player movement, including nimbuses or energy halos around players (Meggyesy 2014).

Meggysey was shifted into this unusual ASC by a recent "dinger," a mild head jolt that, ironically, would have seen him taken out of the game under the NFL's new concussion protocol (in other words, it's not a recommended mode of entering into the Zone experience of clairvoyance).

Telepathy and Psychokinesis

Other psychic abilities displayed by athletes go beyond internal perception and individual performance to affect teammates, audience members, and even physical objects like the ball itself. Auto racer and journalist Denise McCluggage described a basketball game as

a fascinating network of visible energy. . . . Bright cords of varying width connected the Golden State Warriors at their middle. The lines all emanated from Rick Barry . . . [who] was glowing, obviously the hub of the team that night. The changing thicknesses of the cords extending from him indicated where his next pass was going, even when he was looking in another direction. The ball followed a remarkably predictable path down shining corridors of energy (Murphy and White 2011, 135).

As perceived by this clairvoyant journalist, Barry's qi was following his mental intention even when it was being sent outside of his body. His teammates and even the ball itself were connected by this energy. This influence can have a very physical impact and these shining corridors of energy can even be intentionally directed by the players' focus, as the next example will show.

Steph Curry, generally regarded as the NBA's all-time most accurate shooter, recently canned 105 consecutive three-point shots from thirty feet during practice! Upon viewing a video clip of this performance, energy medicine practitioner and clairvoyant Donna Eden described the process to me: "A morphic field can be created with lots of practice. It is like the ball is following an energetic groove."

In other words, the intense mental focus of players like Curry can project energy externally, away from the body, as per TCM's mind/qi dictum. The energy prepaves a route for the basketball to follow on the way to its destination, like the rubber bumpers on the sides of a bowling alley. This pathway in space is invisible to the untrained eye but is perceptible to skilled psychic observers and is evidently sensed unconsciously by athletes as they interact with their teammates and opponents. It is a more concrete example of the mind/matter interactions that were researched in Princeton's random number generator experiments to be described in chapter 8.

The Third Eye: Perceiving the Dimension of Energy

The paranormal side of the Zone was evident in the 2013 baseball World Series, when one player, David Ortiz, had a batting average of .688 (i.e., he successfully got on base with a hit 68.8% of the time), which was more than twice what is usually considered to be a noteworthy average (.300) and was three times as high as the opposing team's average (.224). He was named the MVP (most valuable player) of that series, and part of his success can be attributed to what one clairvoyant described as his "wide-open third eye" (the focus point for clairvoyant perception) and "strong mental focus on his *tan tien*" (the martial arts center of physical energy). In other words, Ortiz didn't need to think in order to perform well—he was already energetically locked into the Zone because of how his chakra energies were being activated (Leskowitz 2014a).

Dr. Laith Mahmood Muhammad Al Azawe (the parapsychologist for the Iraqi Olympic Committee from chapter 3) would assess the energetic status of Iraqi athletes prior to important competitions and prescribe exercises to enhance their energy flow. He was also able to make pregame energy assessments, at a distance, of such factors as concentration, anxiety, and team cohesion, and was surprisingly accurate in his analyses (Al Azawe 2020). A growing literature supports the high degree of reliability of this sort of psychic diagnosis, especially in the medical setting (Colter and Mills 2021), so perhaps energy assessments will someday be a standard part of any athlete's evaluation process.

OTHER EXPERIENCES OF THE ZONE

As previously mentioned, the Zone goes far beyond athletics and encompasses a wide range of creative endeavors. Of course, it's also possible to be in the Zone even while doing mundane activities. As the Zen monk said, "Before enlightenment, chop wood and carry water. After enlightenment, chop wood and carry water." Any activity can be done in an enlightened, fully present manner, and we have certainly received great cultural gifts from nonathletic but creative people who were in the flow state.

Creativity and Performance

Two hundred and eighty years after it was written, George Frideric Handel's masterpiece *The Messiah* remains one of the best-known pieces of classical music. This work for instrumental ensemble and choir takes well over three hours to perform, and yet Handel composed and arranged the entire 259-page score of 53 movements in the compressed time frame of 24 days. He reportedly, "saw all heaven before him" (Luckett 1995), feeling as though he had been given a direct vision of the face of God as his inspiration.

In contrast, composer Franz Joseph Haydn (he of the 106 symphonies) experienced such deep fatigue during his late-life physical decline that he was unable to physically transcribe the musical ideas that continued to "pursue" him (Dies 1810). He described his flow state:

> I must have something to do—usually musical ideas are pursuing me, to the point of torture, I cannot escape them, they stand like walls before me. If it's an allegro that pursues me, my pulse keeps beating faster, I can get no sleep. If it's an adagio, then I notice my pulse beating slowly. My imagination plays on me as if I were a clavier. I am really just a living clavier.

But he had become a clavier (a seventeenth-century French keyboard) that could no longer be played—his etheric body was so depleted of energy by his illness that the connection between mental and physical body was effectively severed. His musical ideas could not be translated into action, leaving him in a mental Zone that could no longer be materialized or downloaded onto the physical plane to be shared with others.

Modern composers experience something similar, as described in a recent essay about songwriter Paul Simon's creative process (Petrusich 2023). "Songwriters often speak about their work as a kind of channeling—the job is to be a steady antenna, prepared to receive strange signals." To do that, Simon would often wake at 2:30 a.m. to jot down musical ideas that were coming to him.

The Zone occurs for writers as well, albeit often going by a different name. Numerous authors and poets have talked about their process of working with their Muse, their source of inspiration (called the "daemon" by the Greeks). In this sense, being "amused" is more than just a lighthearted distraction—it is a connection to one's deepest source of inspiration. Many firsthand accounts of these contacts have been published.

In one musical example, the narrator in the folk song "The Letter," popularized recently by Robert Plant and Alison Krauss, describes to his girlfriend where he went for inspiration when he was struggling to write a difficult breakup letter to her:

> *Please read the letter,*
> *I wrote it in my sleep,*
> *With help and consultation from*
> *The angels of the deep.*

Spirit guides, guardian angels, inner voices, visitations from beyond—a wide range of parapsychological resources can be gateways to inspiration, a key to accessing the Zone of creativity.

Perhaps the most widely discussed nonsports venue for being in the Zone is artistic performance—song, dance, theater, music. But because these events rely so much on performer interactions with the audience and their bandmates, I'll defer a full discussion to chapter 8's look at group energies.

Public Speaking

As I mentioned earlier, my biofield expands measurably when I talk about things I enjoy, and I feel like I'm in a speech Zone. My guess is that great speakers enter this state to an even greater extent and would be found to have much larger biofields while speaking in public—Martin Luther King, John F. Kennedy, Barack Obama, and yes, even Adolph Hitler (he had a powerful enough field to entrance an entire nation, though his energy was pitched at a very low and destructive vibrational frequency—more on this interactive group biofield effect in chapter 8).

FICTIONAL DEPICTIONS OF THE ZONE

Numerous fictional narratives about the Zone offer important insights that have real-life significance and influence on their audiences. I have chosen a few particularly impactful examples to examine in more detail.

Golf in the Kingdom

Many of the vignettes in Michael Murphy's *The Zone* were sent to him by readers of his previous book, the story of a golfing mystic entitled *Golf in the Kingdom* (Murphy 1971). In this fictionalized first-person account, Murphy tells of the life-changing round of golf he played with legendary Scots golfer Shivas Irons. Over the course of one memorable day, Murphy learns how to "ken the world from the inside." The instructions are simple:

> Close your eyes and feel your inner body, a body within a body, sustained by its own energies and delight, a body with a life of its own waiting to bloom.

It's the etheric body, the *pranamaya kosha*, the sheath of prana, as seen by a Scotsman. As Murphy masters this sixth sense, his game improves to the point where he can shoot a hole in one on the fabled thirteenth hold at Burningbush—in the dark! Life imitates art: one of the organizations that Murphy helped to catalyze, International Transformative Practices, includes in its annual fundraising event a blindfolded putting tournament at a local golf course, where participants learn to rely on intuitive perceptions, just as Irons taught Murphy to do.

The Zone went on to provide the intellectual structure for conceptualizing these feats, and the inspired golfing tale that triggered it was named as one of *Sports Illustrated*'s Top 100 Sports Books of All Time. It was the forerunner of the "inner game" genre of sports psychology self-help books, shifting the focus from traditional objective factors like body mechanics and physical training onto the subjective psychological and spiritual aspects of tennis, golf, and basketball. These books have had such intriguing titles as *Way of the Peaceful Warrior* (Millman 1980), *The Rise of Superman* (Kotler 2014), and *The Men on Magic Carpets* (Hawkins 2019).

The Curious Case of Sidd Finch

In 1985, journalist George Plimpton wrote an April Fools' story about baseball phenom Sidd Finch, a previously unknown rookie who took Major League Baseball by storm because he could throw a ball 160 mph—60 mph faster than anyone had ever done before. Sidd (short for Siddhartha, the Buddha's given name, meaning "one who has achieved the siddhis") was supposedly a student of Tibet's great poet-saint Lama Milarepa, and according to Plimpton could create:

> An apparent deflection of the huge forces of the universe into throwing a baseball with bewildering accuracy and speed through the process of siddhi, namely the yogic mastery of mind–body. He mentioned that "The Book of Changes," the I Ching, suggests that all acts (even throwing a baseball) are connected with the highest spiritual yearnings. Utilizing the Tantric principle of body and mind, Finch has decided to pitch baseballs—at least for a while.

Though it was a fictional story (the *I Ching* isn't Tibetan, for starters), it was such a highly evocative example of multidimensional forces being leveraged while in the Zone that fans and sportswriters were caught up in the publicity wave, only to be disappointed when a follow-up press conference was held later in the month to reveal the hoax. Nevertheless, Plimpton had brought these esoteric concepts to a mainstream audience of *Sports Illustrated* readers, and something had clearly resonated at a deep level for millions of readers.

The Queen's Gambit

The Netflix miniseries *The Queen's Gambit* was their most-viewed drama worldwide in 2020, thanks to a Covid lockdown-induced explosion of interest in indoor games like chess. The fictionalized story of an orphan girl in rural 1950s America who rose to fame as a chess prodigy owed its popularity to several factors: topical themes like female empowerment and competition between the United States and Russia, a Golden Globe performance from lead actress Anya Taylor-Joy, and an exciting depiction of playing chess in the Zone. Though chess is obviously not an athletic sport requiring muscular strength and coordination, the Zone of chess involves a mental focus that here verged on the paranormal. Not only could the protagonist mentally review each move in a chess game after the game had ended (a common skill among masters, who can also play a dozen simultaneous games while blindfolded!), she could also *preview* the moves of her upcoming game by watching it unfold like a movie projected on the ceiling above her bed.

Interestingly, Beth's fictional talent mirrors the abilities of the "remote viewers," a class of U.S. intelligence agents who were trained by the CIA to be telepathic spies, as depicted in the movie *The Men Who Stare at Goats* (based on a very true story) (S. Schwartz 2007).

SHORTCUTS TO THE ZONE

Not surprisingly, athletes often cheat to get into the Zone. The pressure to win, and the rewards that accompany it, are so great that many athletes have tried to short-circuit the training process by using artificial catalysts to get in the Zone—drugs. Many substances have been misused to enhance performance, and doping scandals have tainted every major sport. Here is a brief summary of some favored shortcuts into the Zone.

Alcohol: causes disinhibition of the frontal cortex, less self-criticism, and more freedom of expression. Chess Grandmasters Alexander Alekhine and Mikhail Tal had well-known problems with drinking in the latter part of their career, and too many musicians to name have crashed on the rocks of alcohol abuse. Famous artists who created while drunk include abstract impressionists de Kooning and Pollock, as well as Van Gogh with his absinthe.

Opiates: increase bodily comfort. Poet Samuel Taylor Coleridge used opium to access his visions of Xanadu, with its "stately pleasure dome" and "caverns measureless to man." Heroin abuse and addiction has bedeviled musicians from Gilbert and Sullivan's ensemble on up to Billie Holiday, Ray Charles, and whoever is on the cover of this month's issue of *People* magazine.

Steroids: increase physical energy. Too many Olympians to mention have been caught out by drug testing, with steroids being #1 on the list because they massively boost muscle growth, leading to increased strength and endurance. In addition to this physical impact, steroids like testosterone also create an aggressive altered state known as "'roid rage," so strength-oriented sports like weight lifting and football are most rife with its abuse because this aggressive, combative energy can be channeled directly into a more primitive version of the Zone.

Coffee: increases physical energy and mental focus. Johann Sebastian Bach wrote the secular "Coffee Cantata" as an ode to coffee, which he described as "more lovely than a thousand kisses." As an added benefit, coffee also prevented the cantata's main character (a stand-in for Bach?) from "turning into a shriveled-up roast goat," a benefit that would appeal to athletes and nonathletes alike.

Amphetamines ("speed"): dramatically increase attention and energy. Beat poet Jack Kerouac wrote to his friend Allen Ginsberg:

Benny [Benzedrine] has made me see a lot. The process of intensifying awareness naturally leads to an overflow of old notions, and voilà, new material wells up like water forming its proper level, and makes itself evident at the brim of consciousness. Brand new water! (Rasmussen 2011)

A literal flow state. Small wonder that attention deficit disorder (ADD) is so widely diagnosed in professional athletes—a doctor's diagnosis will qualify them for a TUE (therapeutic use exemption) to obtain a prescription for Adderall. It's a medically sanctioned form of amphetamine that is only available by prescription and that has undoubtedly provided legal enhancement to the performance of many elite athletes.

Marijuana: calms the mind/body, enhances perception and transcendental thinking. Visionary artist Alex Grey has become well-known for his iconic paintings of the human energy field that are so precisely detailed that they seem to be photographs of the biofield's subtle anatomy. He has described how marijuana became his favored method of "shifting gears" into a higher state of artistic creativity (Rosner 2002):

> There are times my wife and I will get into a pickle with some particular art work that we're doing, a painting that gets to a certain state of frustration, so we'll get high, usually just smoke a joint, and look at it, and many times the solution is obvious in that state. We're transcending our own conventional modes of thought by shifting gears.

He's not the only one: every rock group ever, for starters. Some health clubs are now providing on-site cannabis access, because its use enhances the workout's potential for full absorption in the moment, a trend started by former NFL star Ricky Williams, with his Power Plant Fitness gym in San Francisco.

Psychedelics: create transcendent and transpersonal experiences and generates inspiration. As Aldous Huxley (echoing William Blake) said about his mescaline experience, "The doors of perception had been cleansed, and I saw life as it truly was—infinite." His 1954 book *The Doors of Perception* was written in and about this exalted ASC that forms the basis for the perennial philosophy that is the core of the world's great religious and mystical traditions. And those were the same doors that inspired the name for the classic rock group The Doors, whose ability to perform in the Zone was legendary.

Magic: allows for limitless possibilities. In the movie version of Bernard Malamud's classic 1952 baseball novel *The Natural*,

Robert Redford confesses that his otherworldly batting prowess stems from his bat's magical powers: "Wonderboy" was created when a bolt of lightning split an ancient oak tree in half, embedding it with supernatural capabilities. This origin story echoes Shivas Irons's "stick like no other," a golf club that was carved from an ancient *shillelagh*, or fighting cudgel. Religious gestures are commonly used by athletes as centering techniques—making the cross, pointing to the sky, kissing the crucifix—to invoke spiritual energies.

Superstition is rife among athletes, whose belief in the power of their rituals is a powerful anxiolytic in and of itself: readjusting batting gloves after each pitch (Nomar Garciaparra), bouncing the tennis ball a dozen times before serving (Novak Djokovic), taking empty-handed practice shots before attempting a free throw (innumerable basketball players), plus lesser-known tactics (Abraham 2022) like cleaning paper cups from the dugout, putting on socks in a certain order, not stepping on the baseline on the way back to the dugout after an inning, and either shaving or growing beards, depending on which way the wind is blowing. Not quite magic, a bit more like the placebo effect.

HOW TO GET THERE

Categorizing and quantifying these physiologic states and indices can be interesting, but it does nothing to help an athlete enter the Zone. That's because these factors are correlates of the flow state and the end products of being in the Zone, but they are not its cause. Opening that causal door has not actually been a specific goal of coaches and sports psychologists until recently. The common assumption has been that flow-state entry is a fairly random event; we can help increase the odds of it happening, but there can never be any guarantees. However, the paradigm is shifting in the world of sports psychology, and what was once taboo is on its way to becoming the cause *du jour*. We'll look now at some of the most effective ways of intentionally creating a Zone experience, in a sequence of methods that span the levels from physical (massage and craniosacral therapy) to cognitive (affirmations and guided imagery) to energetic (EFT, guided imagery, and energy awareness), to pure consciousness (meditation, nondual awareness), with parapsychology (including third eye opening, precognition, and telepathy) playing a surprisingly big role.

- **Physical comfort:** Every sports team now has a massage therapist on staff, to help increase bodily comfort by working out musculoskeletal kinks that inhibit the flow of energy and divert attention from the game at hand. Another body-centered approach gaining in popularity is craniosacral therapy (CST), developed by osteopathic physician John Upledger almost fifty years ago and now championed by former NFL star and CST practitioner Ricky Williams (the same fellow who opened a cannabis gym in San Francisco).

- **Social Conditioning:** Norway has a history of disproportionate success in the winter Olympics. This is due, in part, to their snowy climate. But another key factor is their social conditioning: the focus, at least initially, is on teaching children to enjoy sport; the outcome measures come later. In 2018, the proportion of children regularly taking part in winter sports was put at 93%. Importantly, at a young age the focus is on fun. They do not keep scores for games involving small children, but as they get older there is a huge active talent pool to move up to the *Olympiatoppen* for elite sport training. "We're not many people, but we're a people with passion," said Mons Røisland after earning silver in the men's snowboard big air. The huge emphasis on camaraderie means the team have little room for ego" (Belam and Ingle 2022). The key Zone elements are there—pure enjoyment to learn how to be in the moment and not obsess about outcomes, teamwork to dissipate egotism, plus social support at a national level that builds cognitive and emotional confidence.

- **Guided imagery:** Many studies in sports psychology dating to the 1970s have shown that hours of on-court practice were not the only way by which basketball players could increase the accuracy of their shot-making. Free-throw shooting percentage—taking undistracted shots from the standard fifteen-foot free-throw line—is an easily quantifiable outcome measure. Mental rehearsal of shot-making, using images of successful free-throw shots as the desired outcome, increased accuracy to the same extent as on-court physical practice did (Grouios et al. 1997, 119–138). The body followed the mind's directions.

- **EFT:** Another approach to increased free-throw accuracy uses the EFT acupuncture tapping protocol to release whatever anxiety might be associated with the act of free-throw shooting. In one clinical study, accuracy increased by 21% after a single fifteen-minute EFT practice session (Church 2010). Another research study documented the effectiveness of EFT in improving the accuracy of free kicks in two women's soccer teams in the United Kingdom by a statistically significant amount (Llewellyn-Edwards 2012).

The most impressive story of EFT's sustained impact on sports training comes from the NCAA Division I baseball team of Oregon State University. Their team's psychological consultant, Greg Warburton, has been providing EFT training to team members since 2006, and the Beavers have subsequently appeared in six College World Series (winning two), to become the most victorious collegiate baseball program in the modern era. During a nationwide broadcast on ESPN, their star pitcher was shown self-tapping while waiting on the bench for his turn to take the field (figure 7.4). The TV announcers were clearly puzzled by what he was doing, but conceded "Well, whatever works. . ."

Figure 7.4—EFT in the College World Series of Baseball

Major leaguers also practice EFT, though they're a bit more hesitant to go public with their secrets. They'd lose their competitive advantage, as well as their carefully cultivated image as regular mainstream guys (i.e., not wackos). However, during the filming of *The Joy of Sox*, Red Sox catcher Jarrod Saltalamacchia was kind enough to show me his EFT routine while we sat on the dugout bench before a game (figure 7.5; he discusses his EFT practice at 51:00 of the *Joy of Sox* video). He had learned the technique via online Zoom tutorials, a commonly used teaching method that highlights just how versatile and scalable this simple technique is.

- **Energy awareness:** Chinese martial arts mastery requires years of training to enable the practitioners to become totally aligned with their internal energy flows. For example, aikido teaches cooperative energy-sharing via the well-known mirroring exercise of *push hands*. Karate masters in a state of energetic

Figure 7.5—Red Sox catcher's EFT demo

alignment can survive direct body blows without injury because they can absorb the energy of the incoming strike and literally circulate it throughout the body to dissipate its strength, so that no individual body part suffers the full energetic impact of the blow.

Similarly, Japanese sumo wrestling matches are energy-dependent and are often decided by energy shifts that take place before the first move is ever made. Psychics have observed that when the combatants lock gazes in the face-off before the signal to start, one man's focus will hold steady while the other's will waver. The result has been determined energetically before the umpire has even signaled the start of the physical contest—the man with the more firmly grounded energy field will win.

The Western athletic tradition, in contrast, has not yet embraced this sort of energy awareness. To be fair, practices like yoga are beginning to make inroads, but more often as an aid to flexibility than as a path to energetic balance. Similarly, acupuncture is used for pain relief and recovery from injury, rather than for energy alignment and Zone facilitation. Nevertheless, a growing number of Zone-oriented athletic coaching and training programs and webinars and apps are springing up everywhere (see Resources, p. 274, for info on EVO Sports and other Zone-oriented sites).

Spiritual Practices

So if sport is the yoga of the West, and athletic endeavors become spiritual practices when played from a mindset of expanded consciousness, sport can become as effective a vehicle for unifying body, mind, and spirit as yoga. This vast transformative potential is usually obscured by our Western preoccupation with winning as the most important outcome of a sporting contest. As Shivas Irons said about the quest for breaking par scores on the golf course, "goin' for results like that leads men and cultures and entire worlds astray." In other words, he wouldn't be a fan of the sports analytics movement, but would focus on the intangibles and enjoy the moment rather than analyze it or making winning the goal.

When mindfulness and conscious awareness are infused into sports, the combination can bring athletes directly to the infinite. Paradoxically, because these spiritually oriented approaches require a dropping of the ego, many star athletes are just too narcissistically oriented to be able to take this step (Ford 2022). Nevertheless, here are three representative spiritual practices that have been applied to the world of sport.

Mindfulness

The gap between sports and mysticism is rapidly shrinking, as practices like meditation become mainstream. Meditation was initially touted as a practical technique to enhance athletic performance by increasing attention and focus, but it has also gained favor because it creates a very pleasant ASC. For example, basketball coach George Mumford is a regular on the meditation retreat circuit as he talks of mindful basketball (Mumford 2015), while something as seemingly unspiritual as weight training has been reinvented as a mindfulness practice. Lifting is no longer about listening to distracting music on the earbuds while making your daily quota of reps. Mindful lifting is about consciously attending to the intentional engagement of each muscle with every movement and energizing that process with every breath you take (Tripoli 2016). Herbal remedies are also being used to enhance immersion in this experience, as mentioned earlier regarding marijuana offerings at some gyms.

Crew practice becomes a collective team meditation under the direction of Jim Joy, the rowing coach whose "Joy of Sculling" conferences over the past thirty years have been attended by more than five thousand rowing coaches. The common coaching practice of breaking down each stroke into its individual mechanical elements is transcended by Joy's emphasis on the flow, merging body, mind, and boat so that

> the cyclical stroke is one piece that is not fragmented. There is a strong bond between the rower's body, the shell, and the water; there is a state of flow existing with this integrated whole (Joy 2016).

Joy's teachings exemplify another literal flow state, showing that any sport can be practiced mindfully, and transcendentally.

Nondual Awareness

A term that has gained prominence among spiritual seekers in recent years is "nondual awareness," a state of mind in which there is no distinction between the observer and the observed, between the thought and the thinker. It's when the sense of "I" disappears and all that is experienced is the sense of awareness itself, without the commentary, judgment, and reactivity that are a normal part of our waking life. The term is a translation of an ancient Sanskrit term—*Advaita Vedanta*—that denotes this high spiritual attainment, one whose importance has been stressed by many spiritual masters. But what does it have to do with sports?

Denver-based tennis coach Scott Ford had never heard of Advaita Vedanta when he accidentally stumbled onto an innovative approach to coaching over forty years ago. He found that his own focus and coordination improved greatly when he *stopped* trying to keep his eye on the ball (Ford 2014). He knew it sounded crazy, but when he opened his vision up to include the whole picture—with both a peripheral and a central focus—his game began to flow automatically. He realized that "keeping his eye on the ball" also kept his thinking brain engaged, as it busily figured out trajectories and spins and angles. But when he shifted to an open focus, the inner dialogue stopped and he attained a state where there was no "self as tennis player," just the game playing itself—very reminiscent of Herrigel's "not aiming" Zen approach to archery.

I've had the opportunity to collaborate with Ford over the years and to get on a court with him. As a fair-to-middling high school tennis player who could count on one hand the number of game-action Zone experiences I'd had, I had been politely skeptical of his assurances that his approach could bring anyone into the Zone in five minutes. But he was right—flipping the conceptual/perceptual switch with some simple refocusing images made all the difference, as another part of me seemed to take control of the raquet while I (or what I usually thought of as *I*) stood back and watched.

Ford later became familiar with the teachings of Ken Wilber, the great cataloguer and analyzer of spirituality's multiple dimensions. Ford found that Wilber's language of consciousness training mapped directly onto his tennis training practices, because he was attaining the same sort of nondual awareness on the tennis court that meditators had long sought in the ashram. Wilber, himself a former top athlete, has written about this unexpected application of his teachings into the world of sports (Wilber 2014). Ford's phrase "Getting into the Zone by choice, not by chance" is his way of describing the outcome of his Parallel Mode Process. It turns a once-in-a-lifetime Zone experience into an everyday event, much to the amazement of the traditional coaches who'd long believed that Zone entry was an unpredictable fluke. Not anymore.

Love and Happiness

Positive emotions are closely entwined with flow states, to the point where it can be difficult to separate cause and effect. Is the tail of peak performance wagging the dog of positive emotions, or vice versa? Emerging tennis star Carlos Alcaraz thinks the emotions come first. He "was asked if he smiles so much because of how much he is winning or vice versa. He responded without hesitation: 'I'm winning all the time because I am smiling,' he said. 'And I always said that smiling for me is the key to everything.'" (Carayol 2023).

It's easy to understand how happiness can enhance performance by decreasing anxiety, and the same is true for love. The chess-playing heroine in *The Queen's Gambit* was raised in an orphanage and relied on alcohol and tranquilizers to calm the chronic anxiety that stemmed from her traumatic childhood. Without the drugs, she could not perform well in matches, but she was eventually able to emerge from the protective shell of her withdrawn and dissociated personality when she began to experience the comradeship and support of her chess teammates, and the love of her first boyfriend. She no longer needed the drugs to stay focused during competition because love could now bring her into the Zone, and it did so by the well-documented physiologic process described in chapter 5—heart coherence. To use the wonderful phrasing of Canadian Olympic rowers Robyn Meagher and Jason Dorland, "Love can be your competitive strategy" (Dorland 2018). It's very revealing to note that the Latin roots of "competition" translate to: *com*—meaning "with," and *-petere*—meaning "to strive." Competition becomes a "striving together" rather than a winner-take-all defeat of the enemy; it can bring the energies of collaboration and even unconditional love into the unlikely arena of a sporting contest.

In summary, "all roads lead to Rome," or at least many of them do. We've seen that practices to facilitate Zone entry can be targeted at any level of our multidimensional being—physical, energetic, emotional, mental, and spiritual. By outlining the various elements of these processes, we can see that Murphy was correct in his assessment: sport can definitely be a spiritual path for those who can awaken to its full potential. Someday, athletic competitions will no longer be focused on winning or losing, but on engaging in sport with full consciousness of its transformative potential.

PART IV

Large-Scale
Manifestations
of Life Energy

8
Group Energies

ARE HUMANS LIGHT BULBS, MAGNETS, OR TUNING FORKS?

Did you know, looking at Brad Pitt is a bit like looking at the sun?
Some movie stars have this natural wattage.

ACTOR RAFE SPALL, 2021

For better or worse, I've never met Brad Pitt. The closest I ever came to a celebrity was when I had a selfie taken with Fergie, Lady Sarah Ferguson, at an integrative medicine reception where she was a sponsor. She was charismatic, but she didn't cause my camera's film to overexpose (this was in the predigital era). Even so, the above quote is not just a figure of speech—interesting things happen when two or more biofields interact with each other. By now, you've hopefully accepted the fact that you don't end at your skin, and as an individual human being you are permeated by a field of invisible energy, whether it's electromagnetic, metaphysical, or quantum physical in nature. And interact these biofields do, in very similar ways to the bar magnets you played with as a kid—the ends repel each other if they're both north but attract each other if they're north to south. The bigger the magnet, the further apart they can be placed while still exerting a tangible force. As with magnets, so with people. So let's look at the energetics of these interactions by starting at the one-on-one level, then building up to small groups (like families and teams) and then to big groups (like fans in a stadium, citizens in a country, and humanity on a planet).

Personal Space

Stereotype alert: It can be quite amusing to watch a British person talking to an Italian. They do a sort of dance, as the Italian steps in closer to make his points, while the Brit keeps stepping back and away. What's happening is that these two cultures have very different views on personal space and interpersonal boundaries, and they act accordingly. Obviously, they can see each other and adapt their behavior to what's in front of their eyes, but the same information about how close we are to each other is also transmitted by a magnetic sense that we all have—the ability to perceive our personal space, our aura, our biofield.

In other words, Brits are attuned to repel, Italians to attract (end of stereotyping). People differ in their preference for interpersonal closeness or distance and in their sensitivity to the closeness of others. This sensitivity is an extension of the hand exercise we did in chapter 2, because the entire body is able to pick up these "magnetical" cues of biofield boundaries. How else do we know when someone unseen is approaching us from behind? How did our mothers develop those proverbial "eyes in the back of my head"?

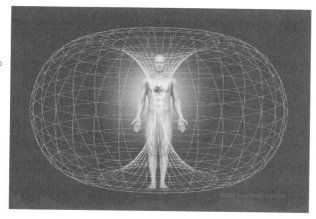

Figure 8.1—The electromagnetic field of the heart

We've all had subjective experiences along these lines to one degree or another, so what does science say? It's not controversial that an electromagnetic field surrounds the body, and the magnetic component of it can be easily measured with a magnetometer, aka a gaussmeter. The findings are summarized in this somewhat stylized illustration (figure 8.1), showing that the edge of the human magnetic field extends five or more feet out into space, with the exact distance depending on the sensitivity of the measuring device and the strength of any given person's biofield.

Magnetic sensing shouldn't be surprising as a plausible explanation of personal space, especially as many nonhuman animals are known to have magnetic sensors—migrating birds that orient to the earth's magnetic field, fish whose lateral lines detect nearby EMFs, and humans with magneto-detecting iron oxide crystals in their pineal glands. And the polarity of magnets—North vs. South—has its analog in personal interactions, as people are repelled or attracted in much the same way.

Charisma

Another way of thinking about my interaction with my yoga teacher's biofield (chapter 3) was in terms of the teacher's personal power, his charisma, because at its core, charisma is an energetic phenomenon. In fact, the term comes from the early Christian Greek *kharisma*, a spiritual gift given by divine grace, a power such as healing or prophecy (Charis was one of Aphrodite's attendants). My yoga teacher had it because he had spent a lifetime building up his reservoir of prana, to the point where it was palpable to others. In a similar way, celebrities like Fergie and Brad have a field, an aura, about them. Famous dancer Rudolf Nureyev had this effect on one of his ballet partners (Crompton 2023): "I swear he had an energetic aura around him. It was a force of nature; it was something that was bigger than him." And while some of that power comes from their own confidence and genius when in the Zone, a big(ger) part is literally projected onto them by their fans, inflating their biofield balloons. It's this fan energy that performers feed off and relish, especially when they are overcompensating for chronic self-doubt—the imposter syndrome that is surprisingly common among well-known performing artists. To a clairvoyant, the energy infusion from fan to popstar is as visible as if they were being sprayed with a water hose. And like any power, this energy of charisma can also be abused, as we'll see later in the chapter when we look at cults.

Tuning Forks

Magnetic fields don't just have a strength or amplitude, but also a frequency—they can be steady, or they can oscillate at a specific vibrational rate. A different metaphor illustrates this variability of the human biofield even more vividly—human beings as tuning forks. Just as a tuning fork will start to vibrate in resonance with a neighboring fork of the same frequency that has been struck, so will people. That's why, and how, emotions are contagious—we literally vibrate in energetic resonance with other people's emotions. The tuning fork image helps us to appreciate the emotional component of biofields, a qualitative aspect distinct from the intensity or strength that the magnet metaphor so clearly conveys. This emotional component is how we pick up the "vibe" when we enter a room, how we're drawn to certain people that we "resonate" with, and how group energies build and grow.

One of the first textbooks in the field of energy medicine was called *Vibrational Medicine* (1988) because it used the language of vibration to describe the energetic aspects of health and healing (Gerber 1996). A modern form of vibrational medicine takes the term literally, using actual tuning forks to identify out-of-resonance areas in the biofield in order to recalibrate them, in a process called biofield tuning (McCusick 2021).

The tuning fork effect actually happens every day, in every interaction we have with one another. I was recently walking the dog in the Covid-deserted downtown business district (all two blocks of it) of our Western Mass village when I came upon one of the shopkeepers standing outside his store, getting some fresh air. We began chatting, but it didn't take me long to realize that he was an unusual fellow. From weather and business, we quickly shifted gears into talking about honesty, the importance of staying true to one's beliefs, the influence of the media in creating divisions between people, and the fact that we're all leaves on the same tree (his metaphor, not mine!). I left feeling uplifted, infused with his hopefulness, and reminded of my own core beliefs—attitudes that had gotten hidden under a cloud of mixed emotions after spending too much time online that morning.

Interestingly, when I tried to recount the conversation to my wife later that day, I couldn't remember the specific examples he'd given, just the overall tone. As Maya Angelou reportedly said, "They may forget what you said—but they will never forget how you made them feel." (In fact, that quote is from a 1971 sermonette by Carl Buehner, the program narrator for the Mormon Tabernacle Choir's radio broadcasts.) Regardless of the quote's source, the significance of that curbside chat was clear to me—we're each tuning forks, capable of vibrating at many different frequencies and needing only a reminder "note" to get back in tune to our truest vibrational state.

And if resonance with one person can be that powerful, then imagine how much more impactful a group's influence can be. That was the focus of a demonstration done by the Institute of HeartMath (IHM), a California-based group of engineers and psychologists who have been studying the magnetic properties of the heart for over forty years.

SMALL GROUP RESONANCE

Since the heart generates the strongest component of the body's magnetic field, it stands to reason that any adjustment in the heartbeat's rate or rhythm or strength would have a major impact on the magnetic component of the biofield. According to the work done by the HeartMath team, that is exactly what happens. The heart is not a metronome, and its rate varies from moment to moment, even from second to second, according

Heart-Rhythm Patterns

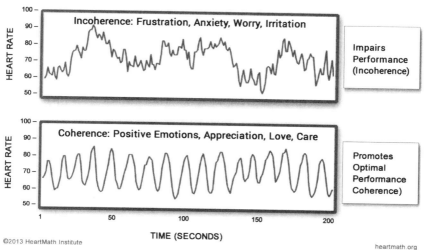

Figure 8.2—Heart coherence,
from the HeartMath Institute/www.heartmath.org

to our breathing pattern and emotional state. As figure 8.2 shows, the heart rate's variability (HRV) differs greatly when we're upset compared to when we're relaxed. In fact, the emotion of appreciation seems to be the most effective trigger for this state of heart coherence, a state of regularly cycling HRV that is marked by stability of the autonomic nervous system and by enhanced mind/body coordination.

I am not one of the thousands of people trained by IHM to elicit this state of heart coherence, and so I was an appropriate guinea pig for a demonstration of group entrainment in their psychophysiology lab. As can be seen in figure 8.3 (next page), I was hooked up to a sensing device that measured my own HRV and heart coherence.

I first established my own baseline level of heart coherence; as Dr. Rollin McCraty, the Director of Research at IHM described it later, I was "pretty incoherent"—not surprising, since I was unfamiliar with their particular heart-centered meditation technique. I was then joined by a group of experienced HeartMath meditators who soon began their internal process after a silent signal from Dr. McCraty. However, I had been blindfolded and fitted with earplugs, so I did not see or hear them

Figure 8.3—Entraining heart coherence

enter the room or receive Dr. McCraty's signal. Nevertheless, I soon felt something shift internally, a lightness in my chest, a pleasant sensation that was actually mirrored by my ongoing heart coherence monitoring.

As figure 8.4 shows, my accumulated heart coherence level took off within seconds of the group starting their heart coherence meditation (the moment marked by the black triangle), quintupling in degree in under two minutes. The demo was stopped soon thereafter, but the results were clear—my nervous system got entrained into the same heart rhythm as the nearby meditators, without any input from the five senses of my conscious awareness. Magnetic field interactions and resonance were at play here, and the implications are surprisingly profound.

For starters, we now have a plausible mechanism for explaining so many intangible human interactions—magnetic fields may be invisible to the eye, but they are palpable to our subtle senses, sometimes even outside the realm of conscious awareness. This demo could be easily fine-tuned to further clarify magnetism's role by using blocking materials to screen out certain portions of the EM spectrum. For example,

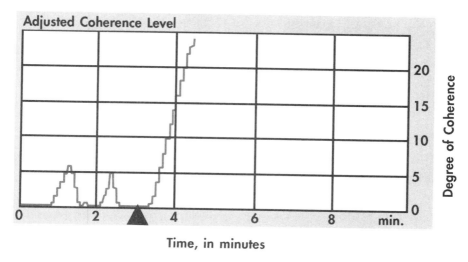

Figure 8.4—Heart coherence results: The black triangle at the bottom is the moment when the team was given the signal to start meditating.

copper mesh screens can shield out the electrical component of the EMF—that's why microwave ovens have an inner mesh lining that keeps those of us "outside the box" safe from the microwaves generated inside the oven. A larger metal screen that encloses an entire person is called a Faraday cage and is used in science museum demonstrations of static electric discharge (figure 8.5), and could also screen the subject of this HeartMath demo from the slightly less scary biofields of the meditators.

In a similar way, a nickel/iron alloy called mu-metal can screen out magnetic fields, which is why it is often used to protect sensitive electronic equipment. So, what would happen if the guinea pig in the IHM demo was shielded with screens made of these two substances? If entrainment no longer occurred, then we'd know the interaction was mediated by electromagnetism: no EMFs, no entrainment. But if resonance still occurred despite the EM screens, then some other force would be at work—lending more weight to the whole question of non-physical etheric energy/qi/prana.

These would be relatively straightforward tests to conduct, but they

Figure 8.5—Inside a Faraday cage

are experiments for another day. In the meanwhile, let's look at the field aspect of some common human interactions, small scale and large, at work and at play, with several simple demonstrations you can do in the safety of your home (unlike figure 8.5!).

Workplace Teams

Burnout is an increasingly common problem among America workers, made worse by the disconnecting pressures of Covid—lockdowns, masks, Zoom, isolation—on top of the disempowering experience of working in corporate settings. One crucial defense against workplace stress has always been the support of your work crew, your team, because when employees feel respected, valued, and appreciated, they are better able to ride out storms. Corporate HR language increasingly uses terms like "join our family" in order to emphasize this cohesion and support, although that image can be a two-edged sword. It's often just empty rhetoric, ironically reflecting the fact that many people grew up in dysfunctional rather than nurturing families.

The ways that team dynamics, especially in families, can affect the

Photo by Kevin Ottalini

group's energy status was analyzed in *The Celestine Prophecy*, James Redfield's New Age bestseller from 1990. One of the key principles laid out in an ancient text discovered by the hero of this fictional adventure story was Insight #4:

> A fundamental assumption underlies most human interactions: we must compete for this energy, drawing it from others and protecting ourselves from others' attempts to draw it from us. . . .We begin to detect our ego's past manipulation devices—control dramas—we have used to build ourselves up at another's expense.

In other words, energy/attention/love is seen by most families as a limited resource in scarce supply, and our adult personality styles reflect the strategies we developed as children to access and hoard this presumably scarce resource. But the next Insight, #5, states that this energy is abundant rather than scarce:

> In fact, competition is unnecessary because subtle energy exists in abundance. . . . Being in a loving state not only connects our energy to the object of our love, but to a *greater source* of energy as well. This is the essence of mystical experience. . . . Love is a background emotion that exists when one is connected to the energy available in the universe, which, of course, is the energy of God.

So when all members of a group—a family, workplace, sports team, or musical group—are aligned to their own personal energy sources, and when they support the others in doing the same, then people can blossom in the larger field of shared energetic abundance. Many new workplace practices in this vein are being implemented by HR departments nationwide, from a moment of mindfulness at the start of a meeting to shared joke-telling to break the ice. But a note of caution is in order: when biofields interact, it's possible to take on other people's energies that are not in alignment with yours. The shared positive intent among team members is a big help in maintaining energetic clarity, but so-called energy hygiene practices provide important ways to release the "stuff" you take on empathically, and are key to self-care and burnout prevention (Leigh 2021). If any HR departments are reluctant to pursue group energy field enhancement, here are two simple demonstrations of the tangible physical impact of the energy generated by coherent group intention and attention.

DEMONSTRATIONS
. . . .
#1 THE SECRET STARER

I've always enjoyed tweaking skeptics and showing them something that stretches their envelope a bit. One such opportunity presented itself at the outpatient clinic where I used to work. At a team meeting that finished early, we had some free time, so I proposed a demo on the power of attention. It was a test that some of my "true believer" colleagues and I had successfully performed at an earlier meeting of our healer's special interest group, so we knew it worked with people who were open to it, but what about skeptics?

We asked the team psychologist (our resident skeptic) if he was willing to be the guinea pig for a painless demonstration of group energy. He assented (he'd have looked like a coward if he said no!), so as the group of us (approximately eight) sat around the meeting room table, we asked him to close his eyes and pay attention to the sensations in his chest. The others would send him positive energy at a silent signal from the group leader and would start and stop these messages at random time intervals. We asked the psychologist to raise his hand when he felt "something," and to lower it when that sensation went away. Much to his surprise (and ours too, to be honest), he raised and lowered his hand at almost exactly the same times that we began and ended our shifts of focus.

It wasn't a carefully controlled experiment, of course, but it'd be easy enough to adapt to any setting and fine tune it. Tighter studies, using blinded subjects, have been done with a related phenomenon: the sense of being stared at (Braud 1993). This is a small-scale example of the type of audience energy that performers the world over love to feed off (more on this later). A video of an energy psychology training workshop led by David Feinstein (*The Joy of Sox* 2013, 49:00) shows a similar group process being activated by a room full of participants, with the muscle strength of the attention-receiver, as measured by kinesiology, being the barometer showing the power of the energy of attention.

#2 THE ROTTEN APPLE

Another simple low-tech experiment can show you how your own heart energy can affect biologic processes. Dr. Leonard Laskow was one of the first physicians to explore energy healing, and he developed a simple demonstration to show how heart energy can be aimed or directed at a target in order to affect natural and biologic processes like entropy—in this case, the decay of an apple—the physical process of entropy (Laskow 1992).

Instructions: Gather a group of people to sit around a table in a healing circle. Cut an apple in half (vertically, so the two halves are mirror images of each other), place one skin-side down in the center of the group, and the other outside the group. Use your favorite healing imagery (a light shining from the heart is universally familiar) to send energy to the apple in the middle of your group. Spend five minutes feeling appreciation for the apple, loving its process of growth and fruition, gratitude for nature's bounty, and the like.

When the meditation is over, place this apple section to the side, out of the group circle but also not next to the other segment. Continue with whatever else was on your group's agenda, and then when the meeting is over, bring the two apple halves together side by side to compare how much browning has happened in the exposed white "meat" of each section. Most commonly, the apple section that was "loved" has not turned as brown as the section that was ignored.

The explanation is that the group energy has buffered the apple against the normal forces of biologic decay. The usual slight browning is a result of oxidation, with the apple's flesh turning color as the oxygen in the air activates an enzyme—polyphenol oxidase—that degrades compounds in the apple to create melanin (surprisingly, that's also the skin pigment that is activated in suntanning to protectively absorb UV light) (McLandsborough 2007). Coating the apple's exposed surface with salt water, lemon juice, or preservatives like sulfite can also block this oxidative process by preventing air, and oxygen, from reaching the tissue. The group's energetic activation of the apple slowed down this decay process, not by keeping oxygen from contacting the apple, but by acting as a biochemical antioxidant that delayed inflammation and degradation, as described in the energy physiology section of chapter 4.

This preservative effect doesn't last forever, of course. As the Buddha noted, "decay is inherent in all component things." Even so, the focused infusion of love-energy can at least slow down the process of entropic decay. A similar but more extreme kind of reverse entropy, in human bodies rather than apples, was demonstrated by the Indian teacher Paramahansa Yogananda. His cross-country lecture tour of America in the 1920s provided our country's first large-scale exposure to yogic philosophy and practice. When he died in 1952, he created another stir when his disciples were allowed to view his body as it lay in state for twenty days before the burial ceremony was held. The mortuary director at Forest Lawn Memorial Park in Los Angeles was amazed at what he observed, and said that:

No physical disintegration was visible in Yogananda's body even twenty days after death. . . . This state of perfect preservation of a body is, so far as we know from mortuary annals, an unparalleled one. . . . Yogananda's body was apparently in a phenomenal state of immutability.

The director's full three-page letter, included in the appendix of later editions of Yogananda's autobiography, is worth reading for its skillful use of professional and technical language to describe the gob-smacking experience of an encounter with the inexplicable. In the yogic view, Yogananda's physical body had spent this lifetime being infused with higher frequency spiritual energies, rendering it literally incorrupt-ible. A bit like the apple but to an infinitely greater degree. And from the materialist POV, "it does not compute."

My most vivid personal experience of group energies creating an effect in the physical world occurred at an energy healing retreat I attended in the late 1980s. Our group of about forty students was learning how to project energy from our hands in sync with our breath-ing. We were sitting in a five-row semicircle facing our teacher, who had us extend our arms outward, with our palms facing toward her. We'd breathe in energy as we inhaled the air and then we projected this energy out from our hands with each exhale. I suspect that none of us would have noticed anything unusual during the exercise, apart from a sense of shared excitement, had it not been for the fact that a large fire was blazing in the fireplace behind our teacher.

Much to our surprise, we saw that the fire began to crackle and flare up as we did the exercise. Then we noticed that the flames rose only when we exhaled and projected the energy out of our hands. When we inhaled, the flames went back down to normal. We began to play with this effect, doing the cycle several times with different time inter-vals, but the result was always the same—our outwardly-directed group energy functioned like a bellows to create a roaring fire. In case any reader's skeptical mind is wondering, the fireplace was a good ten feet behind the teacher, who was herself at least ten feet in front of the first row of students. So even if we had been blowing air out of our mouths as forcefully as possible, we were simply too far away from the fire for our breath to do anything more than make it rustle gently. Yes, Sherlock—it was the energy.

Sports Teams

One of the most widely used idioms to describe coherent group energies comes from the world of sports. "Team Chemistry" is the term for that intangible process by which a team blends, bonds, and transmutes into something greater than the sum of its parts. There's no agreement on which alchemical ingredients create this effect: is it a charismatic manager (Notre Dame's Knute Rockne), an inspiring team captain (the New England Patriots' Tom Brady), a shared religious belief system (pregame team prayer meetings), a time-honored tradition of putting team before individual, or goofy rituals that loop in everyone from star to substitute (singing together, pouring ice water on the coach, high-fiving in the field, etc.)? Regardless of what process any team chooses, the common denominator is the shared energy that builds this team chemistry.

Given how much affection and appreciation athletes have for their teammates, it wouldn't be surprising to learn that they generate the same sort of heart entrainment that I experienced as the HeartMath guinea pig. In fact, some coaches have begun to use heart coherence biofeedback in training teams. Its use with individual athletes is a common training form of biofeedback, but applications to teams as a whole are new (Murphy and White 2011), and the results, though promising, have not yet been published in a peer-reviewed journal. A recent review (McCraty, 2017) describes several other types of human interactions that are marked by high HRV synchrony, ranging from mother/child bonding and friends attuning to one another, to a boy hugging his dog and spectators attending an exciting event. So if this speculation about magnetic field interactions and team bonding is proven to be correct, then "team chemistry" will need to be more accurately relabeled as "team biophysics"!

Despite these common experiences of intangible bonding, our modern digital era seems bent on eroding many of our cherished analog intangibles. For example, the use of statistics to analyze and enhance performance outcomes in baseball was popularized in the 2011 movie *Moneyball,* where Brad Pitt used computer analysis of players' past performance trends and characteristics to assemble a championship-caliber baseball team, despite having the lowest payroll in the league (which meant that he couldn't afford to hire any superstars). The film is based on the true story of the Oakland Athletics baseball team, and it laid the groundwork for what has become the new modus operandi of pro teams—no longer do they rely on the advice of savvy senior scouts, but instead they focus on the algorithms of MIT hot shots. These analysts are members of the Society for American Baseball Research (SABR) and practice the art of sabermetrics. One of their senior advisors, a professor of mathematics at Boston University, was very concise in describing why he and other sabermetricians minimize the importance of intangibles. When I interviewed Prof. Andy Andres for the *Joy of Sox* film, he used the example of Babe Ruth (considered by many to be the greatest player in the history of the game) and said: "Give me Babe Ruth over team chemistry any day of the week. Any day of the week!" He's right, but he's also wrong. He's right in the sense that that an exceptional athlete like Babe Ruth can override many other negative performance factors in his team to reroute their destiny. But he's wrong in implying that team chemistry is not a key ingredient. It's the ultimate intangible and, all other things being equal, it's the little difference that can make a difference.

Musical Groups

Athletes aren't the only people who generate this kind of group energy. The world of musical performance also has a lot to teach us. We've already seen how musicians can get into the same Zone or flow state of peak performance as athletes. But it's not just an individual achievement—their teammates, their band, helps them along.

Consider this vignette from the American improvisational rock band Phish, known for jamming on a single theme for hours at a time. Lead singer Mike Gordon said this about their group process:

> But it takes a certain philosophy of jamming, in allowing the music to be, allowing the group mind to develop and the music to take on its own thing where the individuals aren't controlling it (T. Brown 2011).

Jazz bassist John Funkhouser talked about a similar group mind when he described his trio as an organism (a surprisingly common metaphor among musicians, athletes, poets, and the like):

> The stuff that makes the energy start to happen is friendship and trust and just knowing each other. You can tell a group that's been playing a lot together because they sort of move together and when the music breathes, they breathe. It's almost as though they're a single organism moving together, and it's just an incredibly beautiful thing (*The Joy of Sox* 2013, 21:30).

An exceptionally vivid example of group singing leading to the Zone state is provided by African-American gospel choirs. The interaction among choir members, and between choir and congregation, creates a resonant group entrainment that brings everyone into the state of ecstasy they call being "anointed in the Holy Spirit" (Alexander 2020).

> If you are in the Spirit, you are lifted and you are elevated. And when you get elevated—good God almighty—you're gone! 'Cause God done took over!

That experience includes sensations like "a mysterious joy rising . . . an encompassing warmth or shivering tingling . . . an exceptional lucidity of thought . . . voicing words not their own, uttering wisdom without forethought or intent" (Hinson 2000)—excellent examples of the microphenomenology of subtle energy sensations.

Gospel choirs are legendary for their enthusiasm, the state of having God—*theos*—within them. As an etymological aside, I was disappointed to discover that the word "gospel" is a condensed version of the Old English words for "good story"—from *gōd*, meaning "good" (but widely assumed, mistakenly, to mean "God"), and *spel*, meaning a story or "spell" (i.e., "good tidings"). So, the word "gospel" doesn't actually involve the root "God," as I'd assumed, but it's still divine!

LARGE GROUPS

Mother Nature knows all about team chemistry. We've all watched the flocking behavior of birds and marveled at the amazingly well-coordinated movements of these groups that can range in size from a dozen participants to several thousand. Wildlife cameraman James Crombie has captured striking examples of starling murmurations. The bird-like shape in figure 8.6 soon morphs into other flowing, amoeba-like patterns.

Here is American naturalist Henry Beston's poetic description of those startling starling flocks:

Figure 8.6—A murmuration of starlings

There is no such thing, I may add, as a lead bird or guide . . . . My special interest is rather the instant and synchronous obedience of each speeding body to the new volition. By what means, by what methods of communication does this will so suffuse this living constellation that its dozen or more tiny brains know it and obey it in such an instancy of time? Are we to believe these birds, all of them, are a "machina", as Descartes long ago insisted, mere mechanisms of flesh and bone so exquisitely alike that each cogwheel brain, encountering the same environmental forces, synchronously lets slip the same mechanic ratchet? Or is there some psychic relation among these creatures? Does some current flow through them and between them as they fly? (Beston 1928)

Rupert Sheldrake believes that this "current" flowing through the birds can be explained by his concept of "morphic fields," an invisible organizing principle that gives shape to physical events. The birds' behavior is being coordinated at a higher level of organization than the individual units, by the higher dimensional template of the morphic field. These avian iron filings are moving in alignment with a larger, Phish-like group mind. And yes, schools of fish are another example of coordinated group behavior, and their method of communication is known to be ultimately electromagnetic in nature, as their side markings—the lateral lines—are electrosensitive and can detect changes in local EMFs that nearby fish will be generating. One more example from the animal kingdom: "the spirit of the hive" was the phrase coined by Belgian Nobel Prize winner Maurice Maeterlinck to explain how the organized behavior in the beehive's "inverted city" came about despite having no central point of power like a city hall (Maeterlinck 1913).

Harvard's E. O. Wilson's *superorganism* was the next iteration of this idea, but perhaps the most unexpected version came from the plant kingdom. Trees in a grove are now known to communicate about everything from the weather to pollination via chemicals dispersed through their complex and interweaving root systems and by more subtle electromagnetic signaling methods (Wohlleben 2016).

But the same challenge remains, how is this behavior initiated and orchestrated? Computer modeling doesn't provide a satisfactory explanation for flocking behavior, even after thousands of gigabytes of RAM are used to generate algorithms that map and compare the flight paths of each and every bird in the flock (Zamani 2022). As sophisticated as these digital models are, the attempt to reduce group behavior to a sum of individual movements is a bit like trying to explain how a wheatfield waves in the breeze without mentioning the wind (it's invisible, after all); are we supposed to believe that the individual stalks somehow communicate with each other? These groups—birds, fish, people—move as one analogue organism, not as a collection of digital bits. I'll take Beston over bytes any day of the week!

When it comes to large groups of humans, the examples of emergent effects are quite numerous. Some of these events occur in settings of close physical proximity (musical concerts, sports events) while others happen at such a large distance that nonphysical causes must be invoked (these so-called "nonlocal" effects will be discussed in the final chapter). Some mass events are powerful enough to define an entire cultural era, like Woodstock in 1968. Folksinger and cultural icon Richie Havens spoke of the Woodstock concert as leaving an energetic imprint on that land, in effect creating a Tiller-like conditioned space: "The field holds our hopes, our vibe, our projection. It holds everything. It's sacred ground to us as a generation." (*The Joy of Sox* 2013, 32:30)

Dancing in the Streets: A History of Collective Joy, by sociologist Barbara Ehrenreich, recounts the history of these transformative celebrations, from the Dionysian bacchanals of ancient Greece to the carnivals of medieval Christendom to present-day celebrations like Mardi Gras and Brazil's Carnival. A key element is the feeling of happiness, of joy, that participants experience—it's what nineteenth-century sociologist Émile Durkheim called "collective effervescence." Psychologist Shira Gabriel describes this as something that happens when:

You feel some kind of connection to the other people that are there, even if they're people you don't know. You feel like the moment is special, something that transcends the regularness of normal life, even riding the bus with a bunch of strangers and realizing you're all smiling at the same cute kid (Conrad 2021).

But the forum where I have investigated the energies of large groups most fully has been the sporting event, looking at the energy of the fans in the stadium. And while the five physical senses are obviously deeply engaged in that experience—the sound of cheering, the smell of spilt beer, the sight of thousands of people standing as one—the intangible element is the most intriguing. As is the method used by researchers to study it—the random number generator (RNG).

Measuring Large Group Energies

For over forty years, the Princeton Engineering Anomalies Research (PEAR) lab studied the effect of human attention on random events. Their work sounds like something straight out of the William H. Macy movie *The Cooler*—he played a sad sack loser of a man whose bad luck was so pervasive that he could "cool off" any "hot streaks" at a casino's gambling table just by standing nearby (no spoilers here about what happened to this skill when he fell in love with the casino's cocktail waitress).

That movie was fiction of course, but Dr. Robert Jahn, Chairman of Princeton University's Department of Engineering, found the grain of truth behind this superstition when he created a high-tech version of the roulette table. His team programmed a computer to generate a string of randomly ordered ones and zeroes (so it was more like flipping a coin) at the rate of several thousand times per second—the random number generator (RNG). His team studied how that randomness was affected by human attention, and found, amazingly, that people could skew the output of the computer away from its baseline of a 50/50 distribution of ones and zeroes simply by sitting nearby and willfully intending that the numbers shift. The impact was quite small in any individual trial, but the results became very significant from a statistical point of view when the trials were repeated hundreds and thousands of times. For example, flipping a coin and getting six heads out of ten tries is nothing to write home about, but six hundred out of one thousand would be literally unheard of (try it if you don't believe me!). The usual pattern of RNG output would include occasional brief spikes (runs of ones or zeroes), but only as part of a general "regression to the mean," a trending back to the average 50/50 baseline.

Fan Energy at a Baseball Game

So what would happen if an RNG program was run during a large-scale, highly emotional event like a pro baseball game? A test was set up to see if the output from a laptop computer version of the RNG program would shift away from the baseline 50/50 string of ones and zeroes at any time during a live baseball game and to see what patterns and correspondences with game events might emerge. To do this, I set up my laptop in an undisturbed corner of the bleacher seats in Fenway Park at the start of a Red Sox game. I then moved to a different part of the stadium to watch the game (the RNG program's developer was concerned that if I sat too close to the computer, my emotions would directly impact the computer's output—weird but true), while making notes of events that I thought were particularly significant to the crowd—key hits, moments of anticipation and high anxiety, etc.

After the game ended, I retrieved the computer and sent the output file to the software engineer who had developed this particular RNG program (Scott Wilber of PsyGenics Inc.). He then "decoded" the output and graphed out the ebbs and flows in the data in each successive thirty-second time epoch, without knowing what I had subjectively experienced during the game. His graph would make visible the expected random fluctuations up and down from baseline, those not large enough to be considered statistically significant, as well as any moments where big shifts happened in the RNG pattern. The question

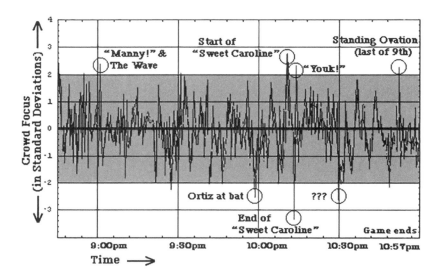

Figure 8.7—RNG output during a Red Sox game

was, how did the timing of these big shifts (if there were any) relate to the timing of the key fan events that I had been subjectively making note of during the game?

The graph, figure 8.7, shows the results, and the explanation follows.

The series of adjacent vertical line segments shows the range of RNG output in any given thirty-second time interval between the 9:00 p.m. start and the end of the game, reading from left to right as the game progressed. The dark horizontal line running across the center of the graph (the X-axis) represents an exact 50/50 balance between ones and zeroes in any given time interval, while any fluctuations of the graph above or below that line but still in the gray zone are interesting but not large enough to have reached statistical significance (<2 standard deviations, as the statisticians would say). However, there were seven moments during the game that spiked above or below the gray band, and so were the moments of true statistical significance. How did these RNG spikes overlap, if at all, with my list of key game events?

As it turned out, of the eight game moments that I thought were most intense (such as a home run by our star player or the crowd starting the rhythmic group cheer called "the wave"), six mapped out *exactly* with those maximal RNG output spikes. The odds of so many correspondences occurring due to pure coincidence were well over 10,000:1, an extreme long shot. In other words, something besides pure chance created this correlation. That something was fan energy, aka collective consciousness, aka the extended biofield of the 35,000 fans. Visionary artist Alex Grey (he of figure 1.7) once brainstormed with me about painting the stadium-wide merged fan biofield that linked everyone together, but the stars didn't align for what would have been a stunning project. So even though we don't yet have a complete explanation for this impact of group mind on computers, it's still demonstrably real.

It's also interesting that the moment of highest crowd impact on my RNG device came when the fans were singing an old Neil Diamond hit called "Sweet Caroline". This tune has become the de facto theme song of the Red Sox, and is traditionally sung during the seventh inning break of every Red Sox home game. It's another example of the power of music as a force that magnifies emotions and increases the power of the biofield. As an aside, here's some related music/sports trivia that is guaranteed to win you a bar bet: "Sweet Caroline" had become so popular with sports fans around the world, including English football (soccer) fans, that BBC Radio listeners voted it to be the official song of the late Queen Elizabeth II's seventieth anniversary Jubilee celebration (Savage 2022). This song is of course in stark contrast to the more solemn music played to commemorate the Queen's subsequent passing, but the deep emotional impact of music in both scenarios serves to exemplify the use of cultural tools as a unifying force capable of generating powerful shared experiences even when separated by great distances (Burton-Hill 2022).

Telekinesis and Baseball

This mind/matter impact is even strong enough to register in the world of concrete physical objects, not just the subtle computer data stream of RNGs. The PEAR Lab used as their macro target the ball cascade box displayed in many science museums to demonstrate the so-called normal distribution pattern of random events. A stream of marbles is dropped down the chute (in the upper left corner of figure 8.8), where it encounters a series of pegs that force it to fall either to the right or the left, down to the next row. At the end of a batch run through a dozen rows of these pegs, the final pattern of accumulated marbles stacked at the bottom of the box shows the bell curve so beloved of statisticians. But the PEAR Lab found that this curve could be shifted in a preselected direction by focused human attention. Again, it moved by a tiny amount that only became evident after summing up the results over many hundreds of accumulated trials, but this was definitely another case of mind over matter in the physics lab.

And so the question arises—if one person can alter the path of dozens of tiny balls, what would be the impact of 35,000 people yearning for a similar shift in the path of just one ball, say, a baseball? A well-known example of this situation happened in the 1975 World Series, when a hard-hit ball heading for foul territory seemed to shift its trajectory in midflight and remain fair long enough to become a home run for local Boston hero Carleton Fisk, much to the joy of the 35,000 hometown fans who were using every ounce of their body

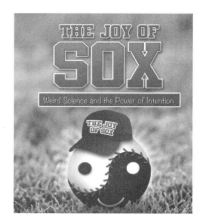

International Consciousness Research Laboratories

Figure 8.8—Influencing the ball cascade   *Figure 8.9—The Joy of Sox*

English and *siddhic* power of intention to affect the ball's path. Of course, this isn't scientific proof of the existence of telekinesis, but every fan at that game was convinced that their efforts helped to create this dramatic result. It's one of the superstitions that keeps bringing fans back to the ballpark, even though most of them are not aware that science actually supports what they're trying to do!

These findings are presented in the film *The Joy of Sox* (figure 8.9), which can be streamed in its entirety for free on YouTube (see Resources).

Regardless of its origins, this fan energy is a force well known to athletes. As Sox outfielder Gabe Kapler said: "It's an indescribable feeling, but the energy definitely comes from the people in the ballpark; it's like the ultimate amphetamine." (*The Joy of Sox* 2013, 7:50)

This energy exchange is a two-way street, because the group energy builds off the interaction between performer and audience. The same dynamic occurs when public speakers inspire huge audiences, with speeches by FDR, Churchill, Martin Luther King Jr., and John F. Kennedy becoming powerful touchstones for an entire nation and generation. But group energies are not all roses and sunshine, as some of the most famous group field effects have come from the dark side of human nature. Some public speakers are notorious for their sinister impact, including master political manipulators like Adolf Hitler, as well as modern-day televangelists, white supremacists, and extremist politicians. The emotions they harness—anger, fear, resentment—can resonate via the same entrainment mechanism that hope and love ride on.

This is possible because tuning forks don't always operate on pleasant frequencies but can span the octaves of the emotional spectrum. In the same way, crowds can resonate to coarse energies as well as to inspiring ones. It depends on the group's intention, the frequency range they are open to, the power of their catalytic leader, and the willingness of participants to engage and activate any specific emotional frequency. For example, when a performer's routine "bombs," he is on a different, nonresonant frequency from his audience. One of the key tools of a cult leader is the use of his (and rarely her) personal charisma to entrain an ever-growing circle of devotees to his particular vibrational mix. The difference between cults and other large groups is that cult members gradually lose the ability to adjust the settings of their emotional tuning forks because they've surrendered that power to the cult leader. In that sense, cult deprogramming becomes another term for resetting the tuning fork, so that people can once again regain control of their biofields.

NO FAN ENERGY

Don't it always seem to go
that you don't know what you've got 'til it's gone.
They paved paradise and put up a parking lot.
JONI MITCHELL, *BIG YELLOW TAXI* (1970)

One unexpected consequence of 2020's Covid-induced societal lockdown was seen in the world of competitive sports. For most of 2020 and 2021, professional sports leagues performed a balancing act by allowing the games to continue (with players training and performing in a variety of isolation "bubbles"), while restricting the number of fans who were allowed to attend the games in person. During the early phase of this rollout, no fans at all were allowed into the stadiums. The Germans called these early matches "ghost games" (*geisterspiele*) because of an atmosphere that was "strangely haunting" and "deeply weird" (Ronay 2020), while aggressive public relations staffers filled in the blanks by inserting cardboard cutouts of fans in the seats and piping in recorded cheers and jeers at appropriate times during the matches. As the year progressed, more fans were allowed in. The February 2021 Super Bowl hosted a crowd of 25,000 people and 30,000 cardboard cutouts (the stadium's capacity was 65,000).

Given all that I've been saying about the importance of fan energy to the game's performance, the Covid tragedy created an unintended experiment, a chance to find out what happens when one of the main ingredients in the athletic experience—fan involvement—is eliminated. How did the players react, how were the games affected, and how was performance impacted when paradise was paved and games had to be played in what was essentially a parking lot? A few answers have emerged, though they're not quite what would be expected.

As was mentioned earlier, the interaction between audience and performers is something that musicians and singers also count on, feed off, and become addicted to. What happens when that nourishment is no longer available? Singer Tori Amos said this about the absence of fan energy during Covid:

> By the third lockdown, I was not at my best. I was grieving not play-ing live for the longest time in my life, not doing what musicians do. There isn't the spiritual ceremony of the collaboration with a live audience. We're talking about a voltage that I can't achieve by myself (Amos 2021).

Though such stories from performers are evocative, we'll remain focused on athletes in this section. I'll present that information a bit differently than in past sections, though. Rather than use narrative discussions, I'll comment on a series of anecdotes and observations made by athletes, coaches, fans, and sports writers as 2020 unfolded. The games that will be described include golf, football (American and British), basketball, snooker (the British variant of billiards), and baseball. I'll even add an analytic research perspective to the situa-tion; though these examples are clearly not controlled experiments, some interesting potential studies are suggested by this point of view. First, we'll see what the overall effect was on performance and the home field advantage (HFA)—this would be analogous to "proof of concept" in the world of medical research. Did the new treatment lead to any detectable clinical impact? Is there any "there" there, as Gertrude Stein might put it? Then we'll break it down ("dismantling," in researcher lingo) to see which component of the fan-less game experience was the key ingredient. To return to medical language, we will examine whether the patients (the athletes) did better because of the medicine (HFA) itself, because they had high hopes (placebo factors), because the doctors spent more time with them, etc. We'll start here with the physical factors—noise levels, visual distractions, and even the possible impact on wind speed on the field. Then we'll look at psychological impacts—players being excited or intimidated by cheers and boos—and then finally the intangible energetic forces that are the main focus of this book, to see how they were affected during the experiment of fan-less sports.

Impact on HFA

Our hypothesis is that the absence of fans had a big impact on outcomes because of its influence on the home field advantage (HFA), the supposed advantage of playing games in the home stadium. As mentioned in the earlier section, HFA is a well-documented phenomenon that is usually attributed to lack of travel stress, ongoing access to home cooking, and the positive emotional support of hometown fans—support that has both a tangible and an intangible element. So, the $64,000 question: What impact did 2020 have on the HFA?

In the case of the professional German football association, Bundesliga, the HFA went missing (Smith 2020):

> The performances of home teams in the Bundesliga have, for all intents and purposes, collapsed in front of empty stands. The number of home victories slipped by 10 percentage points, to 33 percent of matches in empty stadiums from 43 percent in full ones.

Note to American readers: ties are much more common in European football/soccer than in American football, so winning 50% of the matches would represent a huge margin in Europe, while it would be quite average in America. It was the same in America's National Football League, where games almost never end in ties:

Home-field advantage was a myth in 2020. For the season as a whole, the final tally was 49.6% of games won by the home team, compared to 60.2% in 2018 (Volin 2020).

So there was a major drop—10% in both sports. Now let's dismantle the key factors that contributed to this lowered HFA, so we can understand how we got there and what it means.

Physical and Psychological Factors

Distraction: It has become traditional in American basketball leagues for the fans seated behind the basket to try to distract visiting opponents while they are attempting free throws. No fans would mean none of the ever-present hand-waving and banner-shaking distractions. As expected, NBA shooters made more free throw shots during the fanless games after a layoff (when the games had been on hold), testifying to the effectiveness of the fans' pre-Covid efforts (Shaughnessy 2020).
Noise: Professional golf crowds are notable for their hushed silence before each shot, but it's not the volume of the cheers that's crucial. This comment from Tiger Woods shows how the lowered fan presence during the Covid phase decreased player anxiety:

Obviously, the energy is not anywhere near the same (this year). There isn't the same amount of anxiety and pressure and people yelling at you and trying to grab your shirt, a hat off you. This is a very different world we live in (Sullivan 2020).

The lack of distracting noises at close quarters was also credited with the unusually large number of long runs at the August 2020 World Snooker Championship (Guardian Sport 2020). Snooker fans are notoriously raucous in the crowded club atmosphere of the finals, where several hundred fans are all seated within twenty-five yards of the players. Without the fans and their razzing, players were less nervous and performed better.

Wind: An unusual physical fan dynamic was described by Robbie Gould, the place-kicker of San Francisco's NFL football team. He thought it would be more difficult to kick accurately without fans in the seats because wind flow in outdoor stadiums would increase without the surface friction created by the fans' bodies and blow his kicks off course ("the fans usually knock down the wind" Gantt 2020). Gould turned out to be correct—no fans meant more wind—and in fact the only area to decrease in scoring in 2020 were the kickers.

Motivation: The German soccer league commissioned a study of the impact of fan-lessness on the game and found some surprising results. A "negative home advantage" emerged from the stats, with teams performing worse at empty home stadiums than at empty away games as measured by many metrics (goals scored, penalties, possessions, goaltending). There also seemed to be less effort exerted (fewer shots, fewer dribbles), all of which suggested that "the urge to entertain diminishes if there is nobody to respond" (Smith 2020). Also, home teams were penalized more for fouls in empty stadiums than they were when the stands were full, with more fouls over-all being called. This confirmed the earlier finding that much of the HFA arose because referees unconsciously favor the home team and call more penalties against the visitors. The fear of being booed is not restricted to players alone, as it turns out that referees are human, too (Himmelsbach 2020).

Men and women seem to react differently to a missing audience. The lack of spectators had a noticeable effect on the performance of athletes at the 2020 Biathlon World Cup, but with an interesting gender difference. According to one published analysis, when an audience was present, men improved in the conditioning-oriented aspects like skiing (they also run faster), while women performed better in complex coordination tasks like shooting (Heinrich 2021). So, performance anxiety affected men in the strength arena and women in coordination. Conversely, crowd energy seems to enhance each gender's strength—muscle power for men, coordination for woman. It's only a hypothesis, but another one that'd be worth testing.

Intangible Energetic Factors

An eloquent British sports journalist compared fans to a living organism, echoing the musicians we heard from earlier:

> A football crowd is a living, breathing organism. It's a collective enterprise in which individual voices can still be heard. It rises and falls and seethes and sneers and occasionally leaves 10 minutes early to beat the traffic. It doesn't simply react to what it sees; it's an active participant, often scenting a shift in momentum long before it occurs on the pitch (Liew 2020).

He was making the point that the crowd is "more than a sound effect," and even with piped-in cheers, something crucial was still missing. An anonymous online commentator had this telling description (Shaughnessy 2023):

> The irony is the more fake noise and video-game lighting that is pumped into a stadium or park, the more passive the crowd becomes. That stuff is being imposed on the crowd, not allowing them to generate their own energy. Go to any exciting college basketball game and it's ten times more electric-organic crowd energy and participation and the beat of LIVE music, nothing synthetic. THAT kind of atmosphere is much more fun than the synthetic stuff.

Here's how another British writer (also eloquent) described a free-skate routine at U.S. Figure Skating Championship:

> The tension-and-release redoubled with each element (in the performer's routine) as the buzz in the Greensboro Coliseum—that

supersensory communion between performer and audience so dearly missed in the coronavirus era—built towards the climax (Graham 2020).

Leave it to the Brits to capture the elusive with a few carefully chosen words: "a living, breathing organism" and "that supersensory communion between performer and audience." It was this part of the intangible poetry of sports that was eliminated during lockdown.

As we can see, it's hard to tease out the tangible from the intangible factors that are influencing this data. The ideal test would have been to perform the same in-game RNG measurements as I did at Fenway Park, but this time during games with no fans in attendance. Comparing the results from a full stadium to the results from an empty stadium would have been fascinating, but to my knowledge no such trials have yet been done. The closest comparison would be to the off-season and pregame measurements I made at Fenway Park, prior to doing the fan energy test described in the last section.

These were very preliminary statistical "biopsies"—two thirty minute data samples, one taken when the ballpark was empty, in off-season December, and the other taken several hours before the fans began to arrive for the regular season game that I fully analyzed. The data trended in the expected direction—there was more crowd coherence on average during the live game in progress compared to pregame levels, while the pregame levels exceeded those seen in the off-season, which were in turn higher than the levels detected in places of random activity (such as a hospital lobby that I also monitored).

In other words, even an empty, snow-filled Fenway Park had more residual energy coherence than a comparable baseline of random background noise, and that coherence level rose even higher in the hours before a big game, even though the fans hadn't started arriving yet! It was as if the fans' energy of anticipation was already arriving at Fenway hours before they showed up in person. There was a still higher RNG level during the ensuing game, as we've seen, with the highest peaks at moments of greatest emotional impact. The implications of this fact—that a baseball stadium seemed able to store energy even when it was empty—replicates the findings of residual magnetic field changes in a room where energy healings had recently taken place (Moga 2022) and anticipates the next chapter's discussion of "conditioned spaces," sacred sites, and earth energies.

So once again, we have some very intriguing clues, both subjective and objective, that "there's something happening here. What it is ain't exactly clear," as Buffalo Springfield said in 1966. But at least we now have some tools to help separate fact from fiction, to sort out cause and effect.

9
Earth Energies

The force that through the green fuse drives the flower
Drives my green age; that blasts the roots of trees
Is my destroyer.

DYLAN THOMAS, 1934

Hopefully by this stage in our story, you the reader can agree that a good amount of objective evidence supports the existence of an invisible healing life energy. So far, our focus has been on the human aspects of this force as it relates to health and interpersonal connection, but if it's truly a "life" energy, then surely other forms of life in the animal and vegetable kingdoms would also show evidence of its existence, like the earth-sourced plant "force" described above by Dylan Thomas. In fact, many energy-based therapies have been adapted for use with animals, as shown by the acupressure-for-horses chart in figure 9.1. Animal acupuncture is a rapidly growing field with an international professional organization (International Veterinary Acupuncture Society), while Reiki for pets also has a strong foothold (pawprint?) in America today.

This fact shouldn't be surprising, because humans and horses are two branches of the animal kingdom's Tree of Life. As we saw in the phantom leaf discussion, and as is described in both *The Secret Life of Plants* (Tompkins 1973) and the more recent *The Hidden Life of Trees* (Wohlleben 2016), flora also participate in this dance of energy.

EQUINE ACUPRESSURE MERIDIAN CHART

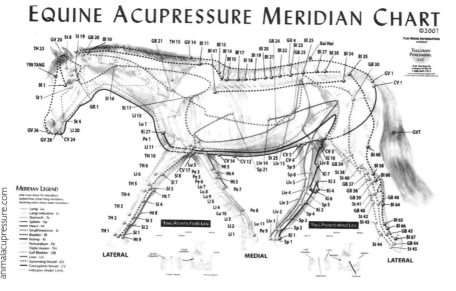

Figure 9.1—The acupuncture meridians of a horse

So, if all three life forms are fueled by life energy, it's fair to ask how far does this feature go? Are rocks alive, or rivers and oceans? What about the Earth itself? Indigenous peoples have for millennia treated Earth as alive and infused with sacred energy, but Westerners have only recently begun to wonder if the Earth itself might be a living organism. Poets everywhere have described this link, encapsulated in the idiom of "Mother Earth," but surely high-tech twenty-first century America is too civilized to take this anthropomorphic figure of speech literally. . .

GAIA AS A LIVING BEING

Science is waking up. The idea of Earth-as-organism was first popularized for contemporary society by environmentalist James Lovelock in 1976 in the form of his "Gaia Hypothesis," which saw the Earth (or Gaia, the Greek goddess of the planet) as an assemblage of interacting life forms: an independent and interdependent self-regulating organism. This view was more refined than commonplace metaphors like "the forests are our lungs and the rivers are our circulatory system," and definitely a big step up from seeing Earth as a "thing" to be owned, to have dominion over, to be divided into political states, and to mine and pollute at will. (More etymology, this time Biblical—the Old Testament phrase "to have dominion over every living thing" is actually a mistranslation of the Hebrew verb *shomer,* which means "to guard or protect," rather than "to control or dominate." Quite an important change of meaning, one that has had a profound impact on Western civilization). But we're still a long way from seeing Gaia as having her own independent consciousness of the sort attributed to her by indigenous tribal peoples the world over.

In this chapter, we'll see that these traditional shamanic beliefs are not so far-fetched and are backed by ancient history, recent science, and an emergent set of hard-to-ignore anomalies that force us to stretch our materialist paradigm past the breaking point. My first exposure to this realm of Gaia and her energies came about because I married an alien (to clarify, she's a very human woman who was born in England, so for many years her official diplomatic status in America was as a "resident alien." But—spoiler alert—I'll also be talking about the other kind of aliens later in this chapter). My marriage gave me the opportunity to visit England fairly often, and I soon became entranced by the stone circles of Stonehenge, the best known of the scores of megalithic structures that cover the British Isles and beyond, even extending to Central Europe and the Middle East (Newman 2017). At first, it was their size and astronomical alignments that caught my fancy, but as I read and travelled more widely, I learned that they weren't randomly sprinkled across the countryside but were following a hidden pattern that could be understood via geomancy, the medieval art of Earth divination.

From my earlier dabblings into the world of acupuncture, I had learned a bit about *feng shui* (literally translated as "mind-water"), the Chinese art of harmony and alignment. It dates back over 3,500 years and strives to harmonize people with their environment through awareness of how landscapes and bodies of water direct the flow of the universal qi through places and structures. Everything from a building's location and structure to interior design and furniture placement can be calibrated to ensure optimum flow of energy through nature and through humans. The practice has become quite popular in America more recently, but most consumers don't know that a European (and perhaps, therefore, less exotic) form of this art was developed in medieval Europe, called geomancy ("earth divination"). That term is the translation of an earlier Arabic practice, *ilm al-raml*, "the science of the sand," a system of charts and symbols drawn on sand and used for a similar kind of energetic divination—yet another reminder of how widespread awareness of earth's consciousness was in earlier times and cultures.

Gaia's Subtle Anatomy

Alignment of human activities with the prevailing patterns of nature's subtle energies was the key to both the ancient Chinese and the medieval British energy arts, but these notions didn't reemerge in British culture until the early 1900s, when retired surveyor and avid hiker Alfred Watkins began to notice that stone circles (as well as holy springs, castles, medieval churches and the like) seemed to be placed in a linear arrangement across the landscape. He called these alignments "leys" (pronounced "lay"), from the Old English *leāh,* meaning "a clearing or meadow," as in "Avonlea." He assumed these lines were shaped by such geographic features as land contours, walking paths and trading routes.

However, the concept of leys developed a more mystical and energy-based overlay when Watkins' 1925 book *The Old Straight Track* enjoyed a resurgence of interest in the '60s. New Agers were drawn to the pagan concept of energy lines traversing the Earth's surface, finding apparent validation in Watkins' maps, much like the *lung mei* or "dragon lines" in ancient China. Watkins' tracks were reconceptualized as energetic pathways that the ancients were attuned to, guiding them to find the optimal placement for their megalithic monuments. This revisioning echoed British occultist Dion Fortune's 1936 novel *The Goat-Foot God,* which described how these "lines of power" were used by medieval witches, as well as by their contemporary heirs, to access megalithic Earth wisdom, a practice which was threatening to the Church hierarchy and to the ruling classes because it was, literally, a form of alternative energy that could potentially undermine the existing power structures.

A revisioning and expansion of Watkins' work is described by Paul Broadhurst in *The Sun and the Serpent*. His extensive fieldwork led him to describe an alignment of sacred sites that extends for several hundred miles across all of southern England, unlike Watkins' tracks that were at most a few miles in length. Broadhurst calls this the St. Michael's line (Broadhurst 1989). This line runs from St. Michael's Mount at the southwest tip of Cornwall through to Cadbury Castle, Glastonbury, Avebury, Sinodum Castle, to Bury St. Edmunds near the southeastern coast of England.

To my mind, this line is in effect a terrestrial acupuncture meridian, with each individual castle or stone formation site functioning as an earth acupuncture point. Some measuring devices, to be discussed in the next section, have produced evidence that tends to confirm this energy-based explanation for the location of the ley lines. This stands in sharp contrast to the nature of England's best-known meridian—the *prime meridian* at the (Greenwich Royal Observatory). This marker for zero longitude was arbitrarily placed in 1884 at the site of the GRO's great Transit Circle telescope, so located because it provided the best vista above London's notorious fogs. The meridian line was actually moved one hundred yards east in 1984 following a global geographic recalibration, highlighting its totally arbitrary and nonenergetically based placement.

So if it's the case that Gaia has her own acupuncture meridians and points, then the next step in subtle anatomy analogizing would be to recast the electrically charged layers of the Earth's atmosphere (the ionosphere and the magnetosphere) as analogs of the planes and subplanes of the biofield, the Earth's "aura." Along these lines, the Global Coherence Initiative has done significant work on the behavioral impact of one of those planes, the magnetosphere (more in chapter 10). And to complete the subtle anatomy comparison, some have suggested that the planet's major sacred sites—Stonehenge, the Pyramids, Notre Dame Cathedral, Ayers Rock, the canyons of Sedona—function as chakras for the entire Earth, setting the energetic frequency range for the entire planet and coordinating the flows of earth energy along the global ley line network (a literal worldwide web!).

Sacred Sites

We've all experienced the sense of heightened energy when we visit a favorite natural setting, and if we're fortunate, it sometimes rises to the level that the Iroquois people called *manitou*, or sacred presence (Mavor and Dix 1989). Many locations around the world have become widely recognized for exuding this universal spiritual "vibe." For example, visitors to the canyons of Sedona, in the American Southwest, are given maps that show the exact location of a series of seven major energy vortices, all in a line. Famed naturalist John Muir's favorite shrine was the Yosemite Valley in northern California. Muir explained that it was easier to feel than to explain its grandeur, saying, "Every natural object is a conductor of divinity, and only by coming into contact with them may we be filled with the Holy Ghost" (Muir 1911).

Other sites have developed their energetic aura following decades of ongoing spiritual and devotional practice by people there, like the Catholic shrine in Lourdes, France. In effect, these shrines have become "conditioned spaces," as described by Stanford physicist William Tiller. In brief, he found that a laboratory room that was used exclusively as the site for regular group meditations underwent an energetic shift in its physical characteristics to the extent that the room itself could alter the pH, the acidity level, of a flask of water placed there overnight. Human intention had conditioned the space to display measurable anomalous behaviors. And if the pH of water can be changed by placement in a conditioned space, then it's not a stretch to imagine that human bodies (which are 65% water) could also undergo significant biochemical changes in such an environment, changes that might manifest as miraculous healings (which would emerge via the mechanisms of energy physiology transduction described in chapter 5).

Closer to home, I have been fortunate to live for the past nine years in a part of Western Massachusetts, the Pioneer Valley, that seems to have a deep spiritual power that exemplifies these ideas. Included among the many significant people and events that have been attracted to the area of Shelburne Falls are:

- The fifty-year peace treaty signed between the Mohawk and Penobscot indigenous tribal peoples in 1708 at Salmon Falls, allowing for hunting and fishing to all who lived within a day's walk of the central waterfall.
- The first American educator for women, Mary Lyon, was born here and founded the first college for women, Mt. Holyoke College, in nearby South Hadley.
- The poet Emily Dickinson also lived nearby, in Amherst, and said this about Nature's spiritual essence:

> *Some keep the Sabbath going to church;*
> *I keep it staying at home,*
> *With a bobolink for a chorister,*
> *And an orchard for a dome.*

- Buddhist shrines in all four major lineages have been built within five miles: the Vajra Hall (Dzogchen), the Vipassana Retreat Center (Hinayana), the Shambhala Meditation Center (Mahayana), and the stone shrine to Dodrupchen (Vajrayana).
- A large Tibetan community has emigrated here, following the designation by Dzogchen teacher Namkhai Norbu that female spirits, *dakinis*, have populated the region under the guidance of their tutelary deity, Goma Devi. Perhaps this creation of a feminine vibration was conducive to the flourishing

of Ms. Lyon, Ms. Dickinson, and the nearby all-female Smith College as well.

- Using a series of complex geometric diagrams, contemporary geomancer Peter Champoux has outlined why he believes that Salmon Falls lies at the very convergence point of the entire Western hemisphere's ley line complex (Champoux 1999).

STONE CIRCLES AND GEOMANCY

This sort of speculation about Gaian energy anatomy and the allure of our favorite natural settings leads to the same empiric questions that face acupuncture researchers. Can these tracks be reliably detected, or are they simply connections chosen among a random array of features in the landscape to confirm the biases (conscious or unconscious) of the mapmaker? It's a reasonable question: perhaps there are so many sacred sites in the UK that a random line drawn anywhere will connect a large number of hotspots (even on the St. Michael's line, several sites are admittedly off track by a hundred yards or more).

According to magnetometers, Geiger counters, and the like, incontrovertible validation of ley lines has not yet been achieved. In contrast, a time-honored but decidedly low-tech technique that geomancers swear by—dowsing—has shown some consistent evidence for the existence of these lines. Dowsing, though widely mocked in the mainstream, is still regularly used by farmers in New England and elsewhere to locate underground springs and ideal sites to drill wells. One study conducted by a team of German physicists in Sri Lanka had a success rate of over 90% in deciding where to drill wells (Popular Mechanics 2004). Dowsing is a variant of muscle testing, the diagnostic process used by many energy psychologists, but with the dowsing rod replacing the muscle group as the visible indicator of the mind/body response to a particular mental focus (that's why muscle testing is also called "ideomotor signaling"—the idea moves the muscle).

But the world of higher-tech measurements does offer some tantalizing tidbits. For example, measurements of magnetic field strength at the Rollright Stones, a circle of seventy-seven limestone boulders in England's Cotswold Hills, show the magnetic field to be markedly weaker within the perimeter of the circle than outside. These results correlated with simultaneous dowsing being done on-site and seemed to show a convergence of external force lines (leys?) at the stone circle, with the stones creating a shielding effect to the external field but creating a magnifying or spiraling effect on the internal fields (Brooker 1983). Others have detected unusual low-frequency bands within the overall electromagnetic spectrogram at Avebury (Wheatley and Taylor 2014).

Amplifying and Disconnecting Energy

These correlations hint at the possibility that the stone circles were built on sites whose special energy coherence could be detected by pagan geomancers, who then designed and built stone structures to amplify terrestrial fields and create an even more highly resonant energetic space, perhaps for the purpose of enhancing and accelerating their spiritual practices. In this same vein, indigenous tribal peoples the world over have used ceremony—human intention, coordinated group behavior, and ritualized actions—to energize and activate Gaia as they consecrate their "mother." By offering prayers to the sacred beings of nature while utilizing group energetic coherence, traditional ceremonies from such diverse cultures as Lakota Sioux (sun dance), Maori (haka), Hopi/Navajo (hoop dancing), and Balinese (trance dancing) have worked with these earth energies for millennia.

This model of energetic coordination certainly aligns with chapter 8's analysis of group energy fields, given the synergy between human group energies and the physical location in which they are generated. It also fills in some rather conspicuous gaps in the mainstream narrative about megalithic structures. For example, a recent (June 2022) British Museum exhibit on Stonehenge was, apart from describing its well-known marking of the solstice and equinox sunrises, peppered with such indefinite and euphemistic phrases as "it is thought," "no one knows," "it is hard to explain," and "one supposes." Very thin gruel for the feature exhibit in the nation's prime cultural repository.

Similarly, the August 2022 Special Issue of the *National Geographic* focused on Stonehenge, also with a point of view that avoided any mention whatsoever of subtle energies. Conventional archeologists and academics are not yet willing to consider the possibility that sites like Stonehenge are energetically alive and were used by locals as catalysts to obtain higher states of consciousness when they worked in direct communion with Gaia and used the stone circles as magnifying lenses. Once again, energy is the missing link in a materialist explanation of human behaviors and events,

In contrast to these ceremonial energy enhancement practices, many earth energy networks have also been dismantled in the course of their history. Sometimes the destruction of the energy network was unintentional, as when the stones lining the great avenue of Avebury were broken into smaller fragments to be used for the more mundane purpose of building cottages in the village. British antiquarian (also a physician and priest) William Stukeley arrived at Avebury in 1719 to find that the breaking up of these stones was already in progress (Smith 2016). Fortunately, he provided many detailed drawings of the site and the stones, including a stunning reconstruction of its original form (figure 9.2), stretching over two miles in length.

The original function of this "Great Stone Serpent" is not known, but it's hard not to imagine an energy amplification system made up of dozens of resonant tuning forks lining the main avenues, directing the energy into the focal points of the central circles, where the village now lies and where the major ceremonies were thought to be held.

On the other hand, the pre-Reformation Catholic Church in England made an organized effort to destroy or disrupt these megalithic structures as a way to dismantle, literally and figuratively, the non-Christian pagan beliefs that were prevalent at the time. Many medieval

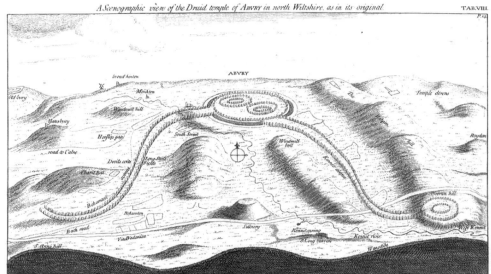

Figure 9.2—The Great Stone Serpent of Avebury

churches were built on sites specifically chosen because they were the home of preexisting stone circles, so that their pagan practices could be more effectively eradicated by "unplugging" their shrines. Consider the example of Stonor House, a British country estate west of London that has been held by the same family since its construction over eight hundred years ago (it was featured in the James Bond film *The Living Daylights* at the 25:00 mark). The family name, Stonor, derives from the ancient stone circle that originally marked the site. When the Stonor family built a Christian chapel on their estate, they repurposed the original megalithic stones to become part of the chapel's foundation, thereby short-circuiting the energy grid.

The administrators of Stonor euphemistically describe this process: "It was frequently the habit in medieval times to 'claim' a pagan circle for Christianity by including a stone in the foundations for the new Christian chapel." At Stonor, these were the blocks of a weather-resistant form of silicated sandstone called *sarsen*, stones that remained behind on Salisbury Plain after the glaciers retreated at the end of the last Ice Age. In another linguistic aside, the word *sarsen* was an adaptation of the Crusaders' Latin term for their Muslim enemies—Saracen—and was used to highlight the English stone's unusual foreign, if not pagan, nature. (What adjectives would megalithic peoples have applied to the Christian chapels?)

Similarly, Mexico City's Templo Mayor Cathedral was built by the Catholic Church on the grounds of the Aztec temple of Huitzilopochtli and Tlaloc. The temple had been destroyed by Cortés and his conquistadors in 1521 and was used for construction material to build the Spanish colonial city there. Alternative theories of humanity's history hold that the destruction of ancient energy focal points was part of a grand plan of the hidden elites to disempower humans across the entire planetary grid, rendering them subservient forever by disconnecting them from the natural energies of the Earth. In one version of this narrative, the 1990 war in Iraq was not about unseating Saddam Hussein or gaining control of oil fields, but about controlling a particularly powerful vortex of feminine energy in that area to ensure that its complementary energies would not threaten the patriarchy's grip on the global economy. Another variant of this theme played out in the hit TV series *Outlander,* where a stone circle in the Scottish Highlands is discovered by the twentieth-century heroine to be an interdimensional portal through which she time-travels from 1945 to 1743 (to rejoin her true love, obviously).

Reactivating Power Spots

If in fact power spots have been inactivated by The Powers That Be over the centuries in order to disconnect their subjects from ready access to Gaia's abundant and innate energy, then it should also be possible to reverse that process. If ley lines are like acupuncture meridians, and sacred sites are like acupuncture points, then there should be an analogue to needle-based acupuncture treatment that could be applied to dormant sacred sites in order to reactivate them. The 1976 rock song "Electrick Gypsies" is just such a call to action for modern energy explorers, as it harkens back to Watkins's old straight track:

> *The gypsy life returns again in a form that's very new,*
> *With motor-vans, electric sounds, and the colours of the rainbow.*
> *We'll open up the old straight tracks and fertilize the earth,*
> *Rediscover the healing sound and ride the psychic surf.*

The same ceremonies that were used by pagan peoples to harness geomantic energy so that they could "ride the psychic surf" could now be used to reactivate and reawaken long-suppressed sites. A fantasy novel from 1985 described how a group of inspired female mystics worked together in this very way to reactivate several important nodal points on the Earth's energy grid and thereby overthrow the nefarious patriarchal Powers That Be (René 2009). Along the same lines but in the realm of nonfiction, I was fortunate enough to participate in one such reactivation process about twenty-five years ago, at a site in southern New Hampshire presumptuously called "America's Stonehenge." In actual appearance, it was a typical New England hardwood/softwood forest with occasional low ledges, scattered boulders, and stone outcroppings, all overgrown with underbrush. Nothing that would have attracted the inexperienced eye, apart from the tourist signs and visitor's center (though to be fair it has recently been upgraded and cleared to highlight the stones).

Archaeological work suggests that the native Algonquians had constructed the site several thousand years ago, possibly for ceremonial purposes, only for it to fall into disuse for several hundred years after the tribes were uprooted by the American colonists. This abandonment led to an energetic stagnation from disuse, one which astrologer Barbara Hand Clow sought to override that day with a group of about fifteen students. She led an afternoon of ceremonial invocations, prayer, sound healing (using toning and conch shell trumpets), and sacred intention to reconnect the granite rocks of the Earth realm with the higher energetic planes of the nonterrestrial beings that she worked with (Clow 1995).

She highlighted several particular rocks as portals, multidimensional doorways through which the indigenous tribal peoples communicated with their ancestors and spirit guides in the higher dimensions. Even as a relatively energy-insensitive participant, I did notice that leaning against several of these particular stone outcroppings generated a sort of elevator effect, as though I were being lifted to an upper story or level of inner awareness. It was nothing as dramatic as *Outlander*, but it was a definite shift in sensation and perception, even though we had not been prepped or prompted beforehand about what we might feel and where we might feel it. So now these multidimensional portals are open again, and as more of them continue to be reanimated globally, Gaia and her inhabitants will benefit from their reconnection to the energies of the higher dimensions.

CROP CIRCLES

Surely one of the most unusual and inspiring photographs of recent years was one taken from a small plane flying over the great stone circle of Stonehenge in the summer of 1996 (figure 9.3). The pilot noticed a formation in the wheat fields that hadn't been there the day before and took a series of photos of the pattern. Similar circular patterns, dubbed "crop circles," had been cropping up (sorry) in the wheat fields of southern England over the past few summers and were beginning to gain public attention, but this one was an order of magnitude more impressive than its predecessors, in both size and complexity. To get a sense of the scale of this image, remember that the circle of stone in the upper part of the photo (i.e., Stonehenge) is one hundred feet across, so the crop formation is almost four hundred feet from top to bottom. The road separating the two halves of the photo is the A435, a busy two-lane highway. Yet despite regular traffic, no travelers reported seeing any activity in the adjacent field during the night before the circle was first discovered.

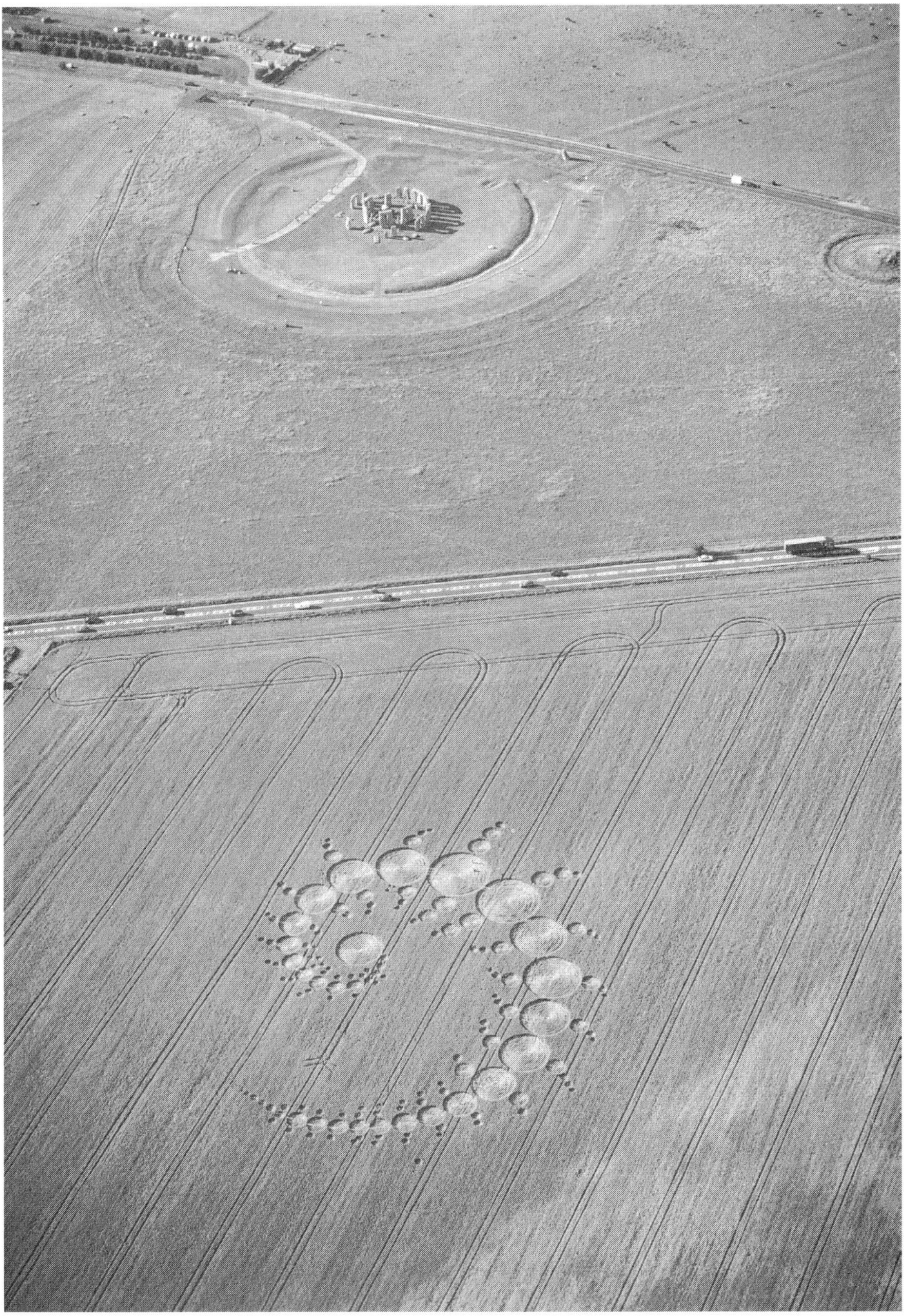

Figure 9.3—Across the street from Stonehenge

Circles like this, though of much simpler design, had been appearing in the British countryside for at least five years. The most widely publicized explanation for the creation of these circles was offered by two retired local farmers in 1991, who said that they used ropes to pull a wooden board along the circular path, while stepping on the plank to crush down the wheat stalks along the route (figure 9.4).

"Doug and Dave," as they were known, became local and national celebrities until skeptics pointed out that they had never given a live demonstration of their technique. A subsequent demo for the BBC in 1999 (BBC) didn't satisfy their doubters because the resultant pattern lacked any degree of complexity and elegance, being a series of lines and circles (as shown in the lower left corner of figure 9.4). And to be fair,

How we made the circles

Today EXCLUSIVE
by GRAHAM BROUGH

THE mystery of the corn circles — which has baffled experts for more than a decade — is today exposed as nothing more than an elaborate hoax by two artists.

After a week-long investigation, we can reveal that Douglas Bower and David Chorley, two men in their 60s, have been successfully fooling the experts for years.

And last night, they destroyed the myths that have built up around the strange circles, which have been appearing in corn fields since the late Seventies.

Under cross-examination, the two men have told a completely consistent story of how they made the circles in fields across the south of England.

Every part of their evidence has stood up to scrutiny. Then TODAY secretly arranged for them to create the ultimate corn circle design in a field in Kent.

But the most damning evidence of all came when self-professed expert Pat Delgado examined the circle and said: "In no way could this be a hoax. This is without doubt the most wonderful moment of my research."

However, just hours before Mr Delgado's visit to the field, we had watched the two men had step by step demonstrated their method of making the corn circles.

Dozens of theories have been expounded to try to explain corn circles. They have ranged from bursts of psychic energy to their creation by UFOs.

But the most "authoritative" voice so far has been Mr Delgado, who has written two books on the subject — Circular Evidence and The Latest Evidence.

Circular Evidence, which chronicles seven years of corn circles in Southern England, reached number three in the best seller list, and concluded that despite the possibility of hoaxes: "There are certain aspects of a single true circle that could never be produced by a machine or manually."

Even the Queen, Prince Philip and Prince Charles have followed Mr Delgado and his colleague Colin Andrews' writings on the subject.

Buckingham Palace has written to him three times. On the last occasion it was Prince Philip wanting to be kept posted on any developments. But

last week, the two men who claim to have perpetrated one of the greatest practical jokes of all time, contacted TODAY.

For seven days, we have questioned them at their homes in Southampton.

ROUND IN CIRCLES: One of the designs Dave and Doug created in a corn field. They always managed to avoid damaging crops and the farmers harvested without any losses.

HOAXERS: How Doug and Dave plotted the corn circles   Pictures: JOHN McLELLAN

All it took was plinths, string

Figure 9.4—True confessions?

the patterns in the English countryside of the early 1990s were composed of straight lines and circles that could conceivably be mapped with plinths and string, but the later ones, like figure 9.3, are as different as calculus is from arithmetic.

For example, the pattern in figure 9.3 is composed of thirty-four separate circles along its main axis, plus over one hundred smaller ones branching out from the sides. The main curve's overall arch is a form of fractal geometry called the logarithmic spiral. This spiral is considered, by students of sacred geometry to be one of the archetypal templates from which the natural world is manifested (Schneider 1994), being present in nautilus shells, sunflower heads, cyclone bands, and even in the arms of spiral galaxies like the Milky Way.

The spiral cannot be produced by compass and straight-edge (D&D's plinth and string), and it would seem unlikely that Doug and Dave knew that its mathematical formula was:

$$x = r \cos \theta = a \cos \theta \, e^{b\theta}$$
$$y = r \sin \theta = a \sin \theta \, e^{b\theta}$$

where *r* is the radius of each turn of the spiral, *a* and *b* are constants that are different for each particular spiral design, *θ* is the angle of rotation as the curve spirals, and *e* is the base of the natural logarithms.

Furthermore, the wheat stalks themselves are intact and pristine, showing no signs of having been trampled upon, whereas the wheat in the BBC demo was clearly squashed. By way of illustration, figure 9.5a is a photograph I took in June 2018 from inside a crop circle that appeared about two miles north of Avebury. The wheat stalks are intact and appear to be individually woven into a precise 3-D structural alignment, with no footprints or signs of trampling nearby. In another circle, the stalks had been bent to a ninety-degree angle at their lowest growth nodes (figure 9.5b) which were elongated and then frozen into their new perpendicular configurations—

Figure 9.5a—Avebury, Summer 2018

Figure 9.5b—Bent but not broken

seamlessly, without a snap or bend, and obviously not trampled by a wooden board.

Incredibly, a sort of energetic imprint persists in the land beneath the crop circles, and it causes the following year's wheat growth to emerge sooner and more robustly in the fields exactly where the crop circles appeared during the previous year, and in the exact same pattern. This growth pattern may be related to magnetic field perturbations within the circles, but their persistence through the year is very hard to explain.

Nevertheless, the legacy of Doug and Dave lives on. Early analyses of the phenomenon included articles like "The end of the 'crop circle' circus," a review in the supposedly objective *Skeptical Inquirer* journal (Baker 1994) that managed to omit any mention of the newer wave of complex patterns. A TV documentary exposé from the late 1990s tried to debunk the phenomenon by showing aerial views of circles that were known to be hoaxes (i.e, made by humans). But, crucially, no ground-view close-ups were included—they would have shown trampled and crushed stalks, as well as a central divot where the compass rod was placed in order to map out the circle boundary. A more recent article in *The Guardian* (UK) newspaper (Myers 2022) focused only on the artistic merit of the circle designs and it prompted over three hundred comments online. Only three of those comments (including one of mine) considered the option that the circles were not man-made hoaxes, but might be the creations of as-yet-unknown forces and intelligences (more on this option to follow). So, the questions remain: how, why, and who? Let's consider the data and the possibilities.

How?

Fortunately this evidence—the wheat—is physical and tangible and so can be analyzed by all the tools of modern science and technology. Building on anecdotal reports that described malfunctioning electrical devices within the crop circle (phones, cameras, laptops, etc.), a wide array of electromagnetic anomalies has been detected in the circles and in the individual stalks of wheat, from increased magnetic field intensities and negative ion fluxes to higher-than-expected levels of background electromagnetic radiation. This is in contrast to the less intense magnetic field measured within the stone circles (Brooker 1983). A more recent summary of crop circle research also includes reports of technical and mechanical failures (cameras, computers) in the vicinity of circles, unusual reactions of animals, disappearance (and occasional worsening) of physical symptoms, and changes in EEG recordings (Pringle 2019).

The actual mechanism that created these circles remains an enigma, with one leading hypothesis being that "nodal swelling" occurs in reaction to "transient high temperatures" within the wheat stalks, thermal surges perhaps caused by microwave radiation (Levengood 1994). Also, piezoelectric effects have been proposed to act through the (possibly) crystalline nature of the wheat matrix (Hein 2002), generating electrical fields near the plants like the crystal radio in chapter 3. Regardless, no mechanical or electrical technology has been able to reproduce the complexity and elegance of these circles. And at least one eyewitness report of a circle being formed (Cowen 2007) makes no mention at all of machinery or equipment, describing only a very brief flash of light illuminating the total darkness of the 3:00 a.m. summer night. The event was caught on video, and the flare was described as "sheet lightning," but the resulting one-thousand-foot-long crop circle pattern only became visible as the dawn light emerged.

So, what does non-mainstream science have to say about this phenomenon? Hypothetical mechanisms invoke torsion field resonance, "dark energy" circuits (Sein), interaction with Schumann waves (Marciak-Kozlowska and Kozlowski 2018) and plasma vortices, but no one has been able to convincingly replicate these structures. Several unconventional approaches like dowsing and intuition have also been used to investigate the nature of the circles, none of which are accepted by mainstream science (echoing the challenges facing ongoing efforts to prove the existence of subtle energy). Two other nonmainstream methods for gaining information about the circles—the use of subjective sensations, and channeling—have led to some additional striking findings.

Subjective Experiences

As described earlier, the science of subjective experience— micro-phenomenology—uses descriptions and self-observations as ways of investigating the world of inner experience. This interoceptive sensory modality delves into the tiniest and most subtle fragments of inner awareness, such as are experienced in deep meditation. This approach is a mirror-image complement to the objective, tech-derived data so esteemed by conventional science. And yet conventional medicine's rejection of the subjective realm ignores the fact that the entire field of mind/body medicine started with this internal point of focus, when meditators began to say, in so many words: "This feels very cool. I wonder if my physiology is shifting when I meditate." That sort of inner examination opened the door to the ongoing flood of neuroscience research about brainwaves, neurotransmitters, immune resilience, and stress-induced illnesses. Both types of research are important, especially because subjective data can often become the launch pad for better-targeted objective studies.

It's the same with crop circle studies—the subjective reports point the way to what the machines should be aiming at. And self-reports of crop circle experiencers run the gamut from inner peace and heightened extrasensory perception to anxiety attacks and physical discomfort. The value of these reports and their validity (that is, their reproducibility and credibility), depends on the skill and training of the experiencer.

For example, I am a long-time meditator and energy practitioner, with at least a rudimentary ability to detect energetic shifts, as validated in prior group trainings and feedback sessions (I've made some progress since my early numbed-out days!). Though I can't identify which specific acupuncture meridian may have been activated in me by any given activity, I can occasionally get chakra-specific bodily feedback along the

Figure 9.6—Meditating inside a crop circle

lines of figure 4.2. And that's what my crop circle experiences have been like, with several aspects of those experiences standing out. One is the sense of awe and bewilderment while walking toward the circles and beginning to grok (a sixties scifi term for "intuit") just how big they really are. Of course, that is in part a cognitive/emotional response to the sensory data that's right in front of my eyes, but it triggers a strong energetic response, including scalp tingling, much like the multidimensional cascade of chapter 5. There's also the experience of being inside a circle to meditate (figure 9.6): it took me only two minutes to reach a depth of inner alignment that usually takes an hour to attain, and that was accompanied by a sense of activation, acceleration, upliftment, and inspiration.

I have not had any classic psychic experiences inside the circles I've entered (as some have reported), but I did experience an unusual synchronicity during one such visit. In chatting with another man who had come into the same circle as my wife and I, we learned that he was also a physician, an endocrinologist from London. And apart from the fact that we were both bearded and balding and Jewish, his expertise in endocrine gland function was an exact mirror to my own interest in chakras; I was interested in their energetic aspects, while he was interested in their physiologic side. We stayed in touch and had a productive clinical and research collaboration over the years as a result of this "chance" meeting.

Channeling

An intriguing point of view about crop circles has been constructed through the process of channeling, the ancient method of accessing information from nonphysical sources. It's too broad a topic to fully address here, but it's a logical outgrowth of the "mental radio" view that consciousness doesn't originate within the brain. To extend the metaphor in a direction never intended by the author of *Mental Radio*, this mental field needn't even be earthbound: formerly human intelligences (i.e., people in between incarnations) and nonhuman forms (extraterrestrials) can also be tapped in this way. Many teachings that have been important for me personally have come in this form: the Seth material in the 1970s, the Alexander newsletters in the '90s, and more recently the messages from Monitor, a collective consciousness rather than an individual being (preferred pronouns: they/them). And despite the common disparagement of channeling as New Age fluff, a growing sense of discernment has led me to trust some, but by no means all, of these sources. Here, then, are some of their important insights into the How and the Why of these *agriglyphs*, as they called the circles.

According to them, the question of where this information comes from is secondary—it's more important that the energetic import of the material be assessed, regardless of its source. In response to my question to the source called Monitor about why crop circles so often occur in southern England, they replied:

Photo by Lucy Pringle

Figure 9.7—Across the street from Avebury

The region has been used for that purpose in recent years because it is as though an artist needs a clean canvas upon which to paint. Most of the Earth does not have that degree of order, as ley lines and vortices [elsewhere] operate with greater variation and irregularity in their designs. The Wiltshire area of southern England has provided an excellent canvas upon which ETs can paint their pictures (Monitor 2018).

Figure 9.7 is another beautiful "painting" on that excellent Wiltshire canvas, in the field across from the stone circle complex of Avebury. Figure 9.2 showed the original arrangement of stones before the village was constructed, and in this aerial photo from the summer of 1994, we can see how the village has taken over the central circle of the Avebury stone serpent, only to be joined by another massive crop circle "across the street."

According to another channeled source named Alexander, these circles have a vibrational function (Stevens 1992):

> Certain new vibrational patterns are being "tried out," as it were, upon the earth's surface. You will note that the patterns grow more sophisticated and elegant with each passing year, your clue that these new vibrational patterns are growing in sophistication and strength, and that after this experimental period they can spread to change the vibrational field of the entire planet.

The changing vibrational field that Alexander was referring to is related in part to our culture's view of extraterrestrials. Given that no adequate human-based explanation has emerged for the circles, we are being forced to consider something improbable, à la Sherlock Holmes: these circles may represent the tip of the ET iceberg. But with an important difference: any off-planet civilization that could produce such amazingly complex imprints has technologies that clearly dwarf ours, which means that they could also defeat and enslave us if that was their true motive. In fact, if the circle-makers behaved like 99% of the aliens in Hollywood films, they'd have conquered us long ago. So the logical conclusion we must consider is the possibility that, in a total reversal of the media's negative portrayals, ETs may instead be here to inspire and guide us. What if the circles are another step toward disclosure of an ET/human partnership, the step that dissolves our media-induced fear response to ETs by exposing us to example after example of inexplicable beauty and awe? From this perspective, the circles may be a key step in the erosion process by which our culture's adversarial and militaristic approach to life finally loses its upper hand, and we allow ourselves to consider new and wonderful possibilities about life on Earth.

In another provocative response along these lines, Alexander poked gentle fun at the human need to seek a logical explanation for anything and everything. By way of context, some researchers have sought to decode the crop patterns by finding circuit diagrams or Tesla blueprints embedded in the circles (Braden 1997). But instead of this, Alexander challenges us to enjoy the simple sense of wonder provoked by the circles:

> The public is enraptured, puzzled, mystified; while the deans of science are apoplectic. . . . As you stand on the brink of a new world order you need such phenomena to remind you of the immense complexity of your universe and how little you know of it; you need phenomena which bring delight and wonder and mystery and confound the guardians of scientific dogma; you need patterns of striking beauty and sophistication appearing overnight in corn fields to reawaken your child's wonder at life. Your reactions are an essential element of the larger phenomenon, which any "rational" explanation would strip away. For us—for anyone—to offer such a definitive explanation would therefore bleed the crop circles of their power over imagination and rob them of their meaning.

They're asking us to follow the path suggested, in much simpler terms, by folk singer Iris DeMent when she confronted the existential puzzle of where we come from and where we go. Rather than work it all out intellectually, she decided: "I think I'll just. . . let the mystery be."

10
Global
Consciousness

Any man's death diminishes me,
because I am involved in mankind, and
therefore never send to know for whom
the bell tolls; it tolls for thee.

JOHN DONNE,
NO MAN IS AN ISLAND (1624)

The take-home point from the first nine chapters of this book has been the idea that each individual person, each group of people, and even Gaia herself all partake of the flow of universal life energy. But if we're all swimming in this same ocean, so to speak, then we might wonder how far the ripples travel—we might even be connected at a higher and deeper level that doesn't even require physical proximity to operate. That's been hinted at in earlier chapters, especially with the mental radio metaphor and the nonlocal aspects of mind/matter interactions. But this final chapter will make the great leap beyond energy and into consciousness itself, the most subtle plane of the entire energy spectrum. The main jumping off points will be the rather diverse springboards of Princeton University (again), magnets (again), the Beatles, a Jesuit paleontologist, and *Hair* (the musical).

THE GLOBAL CONSCIOUSNESS PROJECT

As mentioned earlier, researchers at Princeton University's PEAR Lab have been putting numbers to group consciousness phenomena for over thirty-five years. Once the PEAR team was able to miniaturize the measurement system and program it into a portable laptop, it was a simple matter to measure events outside the lab, like the Red Sox ballgame and the Burning Man festival (Delorme et al. 2017). The next step was to place many such computers at affiliated labs around the world, and that's what they did in a parallel endeavor directed by Roger Nelson, the PEAR lab's coordinator of experimental work who became the director of the Global Consciousness Project (GCP).

Since 1985, over sixty of these random number generators have been recording their output continuously, 24/7, 365 days a year. The hope was that at certain points in time, presumably during globally important events, there would be RNG correlations across this array of devices even though they are widely separated in space. Their entire database is posted on the Global Consciousness Project's website, named after the French Jesuit paleontologist and priest, Père Teilhard de Chardin. In 1955, he popularized the notion of a *noösphere* (meaning a "mind sphere" of collective human awareness), and of an Omega Point (a positive inflection point that would bring us to the next stage in the evolution of human consciousness) (de Chardin 1955). The GCP website's subtitle captures the essence of their quest to find hidden patterns: "Meaningful Correlations in Random Data."

So, for any event of larger global significance, it is now possible to check with the database to see what the laptops were registering at that moment in time. And over the thirty-five years of the project's existence, correlations have been established across a wide array of events at a probability level of 1:1,000,000,000. In other words, the odds are vanishingly small that these results and correspondences occurred as a result of random chance. The labs are too far apart to be influenced by

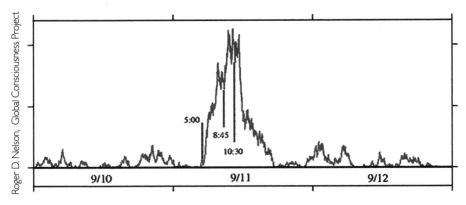

Figure 10.1—The GCP on 9/11

biofields because those don't extend for hundreds of miles, but something is connecting the devices, even if we're not yet clear what the mechanism might be (sound familiar?).

The GCP has tracked many events that commanded worldwide attention, from Princess Di's funeral to athletic championships and natural disasters. But the results that garnered the most media attention, because they were so astonishing, came from the 9/11 attack on the World Trade Center. That day's spike in global coherence was one of the highest ever recorded, and unmistakable even to a skeptic (figure 10.1). But what makes this graph even more stunning is the fact that the RNG peak began two hours *before* the first plane hit! It's as if there were a sort of global precognitive process in action that somehow registered on the GCP's RNG network, the global One Mind that Larry Dossey and others have written about (Dossey 2014).

PASSIVE VS. ACTIVE INTERCONNECTIONS

The GCP focuses on passively detecting the impact of distant events on their RNG network, but other researchers have focused on intentionally influencing distant targets, ranging from RNGs to living systems. From the first reports showing a clinical impact of distant healing prayer (intercessory prayer) (Byrd 1988), the study of nonlocal effects has benefited from the application of the scientific method to an implausible and seemingly impossible phenomenon. Several nonclinical forms of research have also been initiated in an internet-based online format that makes larger-scale participation possible.

Rupert Sheldrake (of morphic fields) and journalist Lynne McTaggart have each recruited participants from around the world to focus their remote attention on specified targets at predetermined times—objects ranging from plant growth (like Grad's original studies) to world peace (like the Transcendental Meditation study). Their results are quite promising and can be reviewed on their respective websites. Of course sports fans use their remote attention to do this all the time when they watch their favorite teams and players on TV—they're rooting, but in an atomized, uncoordinated, energetically unsophisticated and nonlocal way. Which raises an interesting question: what would happen if a selected group of fans was trained in an energy enhancement technique and then given specific targeted events and players to focus on?

I've taken part in two such trials. One was a behind-the-scenes gathering of five energy healers who were invited to attend a professional sporting event at a certain undisclosed location (which I cannot specify due to our informal NDA, but which the discriminating reader can probably guess), in the hope that we could use our energy projection skills to influence the outcome of the game. As much as we enjoyed our excellent box seats (courtesy of management), we were not able to affect the game in any noticeable way. Unfortunately, we did not spend much time coordinating our efforts within the group or deciding what our particular goals or techniques might be. Because each of us chose their own approach to projecting and directing energy, we did not tap into the potential benefits of coherent group intention and coordinated output. We never got a chance to refine our protocol because, understandably, we were not invited back. We have since heard many reports, not for attribution, that other pro sports teams have used this approach, but none have gone public yet (fear of ridicule plus losing competitive advantage).

The other event took place during the production of *The Joy of Sox,* when my own weekly blog postings about the power of fan energy took on a new urgency because the Red Sox were mired in a deep early season slump. Specifically, their number one slugger, David Ortiz, had gone nearly a month without hitting a home run, the longest drought of his career. So I notified the followers of my blog, about eight hundred strong, that when Ortiz came to bat for the first time during that night's game, we were to send him positive energy via the HeartMath appreciation protocol described earlier. That evening, my wife and I went to dinner at a local sports bar with some friends, and I made sure that I was facing the TV. I explained to my tablemates what I had been setting up with our JoS followers, so we took a break from eating to watch Ortiz's first at bat. You could have knocked me over with a feather because, on the second pitch, he hit one out of the park. "Damn, this stuff really works!" was all I could say. Of course, this could have been just a coincidence, but that's exactly why organized studies like the GCP are so important, to move past individual events, anecdotes, and occasional coincidences to uncover true causal connections.

MAGNETIC INTERCONNECTIONS

One established physical force could account, at least in part, for these distant, nonlocal effects, and that is the magnetic field of the Earth as a whole. Gaia is much like a bar magnet, albeit a round one, with lines of force connecting the north and south poles and encircling her on all sides. Every living creature on her surface, no matter where they live, is affected by any ebbs and flows in the earth's overall magnetic field. And ebb and flow it does, with fluxes being generated by everything from solar storms and the phases of the moon to terrestrial thunderstorms and shifts in the location and strength of the poles themselves. So at least from the perspective of our own personal magnetic fields, we are literally like those oft-mentioned iron filings, being influenced by, and participating in, the greater Gaian field. Two aspects of this magnetic interaction are important to mention: a local effect that impacts our personal biologic health, and a nonlocal aspect that maintains our sense of global human interconnectedness.

Despite its offputtingly grandiose title, the 2010 book *Earthing: The Most Important Health Discovery Ever*! shows the importance of each human's direct connection to the earth's magnetic field—through the soles of our feet (Ober, Sinatra, and Zucker 2010). Rubber-soled shoes, linoleum tiling, living and working in tall buildings—all these factors break the circuit of electron flow between earth and body by insulating and unplugging us from the earth's field, causing an unhealthy build-up of charge in our bodies that leads to chronic low-level inflammation and disease. Many physiologic measures (heart rate, blood viscosity, skin conductance) can be enhanced by such simple reconnective measures as walking barefoot outdoors in nature—*earthing*—or by using electrically conductive and grounded pads while indoors—*grounding* (Oschman 2015). Furthermore, clairvoyant assessment shows that etheric energy from the earth can be directly absorbed through the minor chakras at the soles of the feet, so both electrons and prana get to flow with these practices (which means that walking barefoot at Stonehenge would be the ultimate energy booster!).

This individual or personal connection to the greater electro-magnetic field of the Earth opens the door to the next level of intercon-nection, one in which the Earth's field is seen as the conduit through which all humans can interconnect with each other.

Global Coherence Initiative (GCI)

Researchers at the HeartMath Institute have expanded their work beyond clinical trainings to develop heart coherence for physical and emotional health, to link individual biofields to the larger magnetic field of the Earth, in what they call the Global Coherence Initiative (GCI). Their primary insight is that many human physiologic rhythms and collective behaviors are synchronized with, and affected by, the Earth's geomagnetic activity. Our nervous system is coupled with these external influences by resonance, as certain frequencies of rhythmic human function (like EEG brain waves) occur in the same range as the frequency of the earth ionosphere's resonance (the Schumann resonance, about 8.5 cycles per second—the rate at which electromagnetic signals travel around the globe within the containment vessel of the ionosphere).

The GCI hypothesizes that "large numbers of people, when in a heart-coherent state and holding a shared intention, can encode information onto the earth's energetic and geomagnetic fields, which act as carrier waves of this physiologically patterned and relevant information" (McCraty 2018). In other words, it's a two-way street—as more people learn to meditate and attune their biofields to the frequencies of peace and compassion, the earth's field will resonate to this state of increased coherence and thereby spread this emotionally healthy imprint around the globe, even to others who may not be meditators.

The GCI has placed a network of sensitive magnetometers around the world, much like the GCP, for ongoing monitoring and research. In one study, they found that "the applied Heart Lock-In meditation technique has a positive impact on the synchronization between the human heart rate variability and the Earth's magnetic field." This gives new meaning to the term "interconnectedness" (McCraty et al. 2018), which is now not just a social phenomenon but a matter of classical physics and electromagnetism.

THE BEATLES

We've already looked at the powerful energetic impact of music—on the individual performer and on the audience, even in nonmusical gatherings (i.e., "Sweet Caroline"). The global impact of music is embodied in the Playing for Change website, which hosts dozens of videos that are renditions of popular songs performed by musicians around the world but recorded as a single synchronized, sequential, composite performance. Go to their website to experience the inspiring sense of global interconnection that tens of millions of their viewers have felt when listening to their wide range of multi-instrumental performances—music as an international language.

This transnational impact of great music was seen in Paris during their mass unity rallies of 2015, when a balcony boom box recording of John Lennon's song "Imagine" spread in an organic wave of spontaneous song through the huge crowd gathered at the Champs-Élysées. More recently, a cover band for the '70s rock group Chicago played before sold-out crowds on its 2022 American tour, not only because they are excellent musicians and Chicago's music is still popular, but most importantly because the musicians come from Ukraine and Russia. The audiences for Leonid and Friends are inspired by the crystal-clear symbolism of their collaboration in the midst of war, even before they play their first note. In a way, their eloquent opening comments about the unifying power of music were redundant—the audience got the message just by being there. I was fortunate enough to attend their final 2022 US concert—the RNG output would have been off the charts.

One musical event that was monitored by the GCP was the World Sound Healing Day. An annual event organized by pioneering sound healer Jonathan Goldman (Goldman and Goldman 2017), participants around the world are invited to sing/tone/chant/hum their heart-centered good vibes (literally) in a communal prayer for humanity and the earth. As Goldman describes it (Goldman 2023):

Ancient mystics and modern physicists share an awareness that everything in the universe is in a state of vibration. Intentionalized vocal-sound vibrations that are created with the conscious intention of compassion and appreciation can be used as energy medicine for both personal and planetary healing.

This self-created sound—a hum, or Om, or Ahhh will magnify even further the electromagnetic field that develops when our heart and our brain are in a state of coherence. And that is one of the reasons why the different prayers on our planet are vocalized—they are chanted, whispered, spoken or sung so that sound can focus and amplify the power of prayer and meditation. When groups of people synchronize together, sending intentionalized sound for a specific purpose—as on World Sound Healing Day—it brings peace and compassion to our planet, as demonstrated by research from the Global Consciousness Project (figure 10.2).

A metaphysical explanation for these examples of higher dimensional musical/emotional resonance was eventually developed as an outgrowth of explorations in consciousness undertaken by those leading edge explorers of the '60s—the Beatles. In the last few years of their time together, the Beatles began to explore Indian music and spirituality, largely through the influence of George. The sitar solo in "Norwegian Wood" (1965) was a first for pop music and helped George's sitar teacher, Ravi Shankar, became a key bridge figure to the West. Yoga philosophy was touched on in songs like "Within You and Without You," and George's meditation teacher, the Maharishi Mahesh Yogi, played a similar role in introducing Westerners to Eastern mysticism. When the Beatles traveled to India in 1968 to study Transcendental Meditation (TM) with the Maharishi, it was an iconic cultural event that proved to be the thin end of the wedge (as the British say), opening the door for the mindfulness surge in twenty-first century American culture. TM at its peak had dozens of centers in America, several million practitioners worldwide, and an accredited American university, Maharishi International University (MIU), in Fairfield, Iowa. MIU had the foresight to begin performing scientific studies on the nature of meditation and its impact on human functioning in the 1970s (Williams 2020), studies which were instrumental in advancing the field of mind/body medicine and self-regulation, and in exploring nonlocal aspects of consciousness.

Many of their studies were fairly conventional—the effects of TM on blood pressure, the release of stress hormones, and brainwave function. Other claims were controversial and not well documented. For example, they claimed that regular TM practice could promote the emergence of the *siddhis*, the yogic superpowers that Michael Murphy described in *The Zone* (chapter 7). Most notoriously, the TM people maintained that advanced practitioners could levitate via "yogic flying." The term conjures up images of meditators rising up in their sitting lotus posture to float several feet above the ground. However, the available video clips of this miraculous event only show a sort of seated hopping across the cushioned floor, and so these claims are generally taken with a very large grain of salt.

However, MIU's work on group consciousness was more comprehensive and has left a mark on the scientific literature. The Maharishi taught that humans participate in a Unified Field of Consciousness (i.e., the noösphere) that each individual mind can access, so TM practitioners were strong believers that the interactive effects of meditation could be mediated and propagated by this field. Hence, large group gatherings were encouraged in order to amplify the impact of meditation, and daily group meditations of the students and faculty in the MIU amphitheater were a normal occurrence. Anyone who has ever attended a meditation retreat (TM or otherwise) can attest to the power of this group phenomenon—a similar magnification occurs in concerts, ballgames, and the like, but the meditation experience is intensified by a silence that facilitates awareness of the nonphysical aspects of the process. Not surprisingly, measurements have shown increased HRV, heart coherence, and magnetic field strength during such retreats.

A corollary of this transcendent worldview was the so-called Maharishi Effect: if enough people in a particular region (a room, a city, a planet) attained a high enough state of consciousness during meditation, the ripple effects would have an impact on everyone in the vicinity, even calming the minds of nearby nonmeditators. MIU physicists calculated that a subgroup comprising less than 1% of the population could be an effective catalyst to trigger the Maharishi Effect (actually, the square root of 1% would suffice, meaning that a city of one million people could theoretically be transformed by one hundred people meditating together at the same time).

Effect of group meditation on crime rate

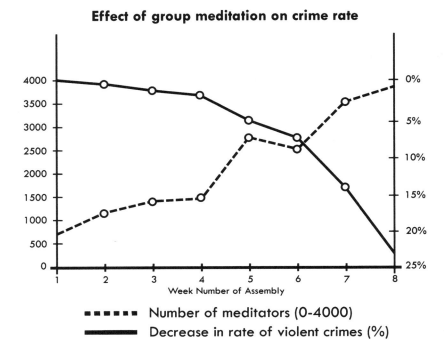

Figure 10.2—The effect of TM on the crime rate

To their credit, the TM researchers set up a study to test this unlikely hypothesis. They arranged for a large meditation conference to take place in the city of Washington, DC, for eight weeks in the summer of 1993, and during that time interval they monitored the weekly violent crime statistics that were routinely compiled by the metropolitan DC police department (which was not aware of the study in progress). The number of meditators grew from an initial 600 at the start of the retreat to a final count of over 3500, and during that time the crime rate (specifically assaults, rapes, and homicides) went down by 23% in greater DC, as shown in figure 10.2. Details of the methodology and statistics are described in an article in a peer-reviewed sociology journal (Hagelin et al. 1999, 153–201), showing that these results could only be explained by a causal connection between two seemingly unrelated occurrences—group meditation and the local crime rate.

These results were so impressive that TM developed a political arm, the Natural Law Party, as part of their effort to expand these practices nationally. They even fielded a candidate in the 1992 presidential elections, the same Harvard-trained quantum physicist who directed the TM/crime study, John Hagelin, PhD. His name appeared on the ballot in twenty-eight states that year, and in the next election cycle, in 1996, he received over 113,000 votes nationally. This total is quite a bit larger than the square root of 1% of the total number of voters across the USA (only about three thousand people), so apparently it's harder to trigger the Maharishi Effect in the nitty-gritty world of electoral politics than in the refined setting of a meditation retreat.

THE COVID-19 PANDEMIC

The final example of interactive large-group behavior we'll look at is the definitive event of the twenty-first century—the Covid pandemic. Its impact has been so deep and far-reaching that it will take years to completely understand its full implications. In this section, we'll consider one aspect of it, a conceptual one—how our culture's choice of metaphors with which to interpret the pandemic set up widely divergent future scenarios. The three most widespread Covid narratives maintain that: 1) we're being invaded by a common enemy (the virus); 2) we've been brainwashed into groupthink via the media-industrial complex; or 3) we're co-creating a mass event through our shared power of intention. These viewpoints each engender a sense of unity and interconnection, but only one includes a recognition of the active, creative, and global nature of consciousness that we've been outlining. Here's a closer look at each one in turn.

1. Epidemics Spread by Germs
The conventional medical view of Covid is that it's a viral illness whose spread can be stopped with vaccines and public health measures that prevent the spread of the virus particles, like masks and lockdowns.

As I mentioned in chapter 4, this germ theory of disease has been the mainstay of the medical model of health and illness since it got a jump start with Paul de Kruif's 1926 bestseller, *The Microbe Hunters*, a book that inspired thousands of young Americans to pursue medical careers. The book's thrust—heroic men used their brains to find the "magic bullet" that would outwit their tiny bacterial enemies and end scourges like malaria and polio—set the stage and defined the terms for our current battle against the Covid-19 virus. As an exemplar of this process of cultural inspiration and imprinting, one of those inspired American children was my father, whose life experiences highlight how our views about epidemics have shifted in the ninety-plus years since he was a child.

Born in New York City in 1922 to Jewish immigrants from Russia who had fled the Czar's pogroms, his pantheon of childhood heroes was topped by Babe Ruth, Beethoven, and Louis Pasteur. He and his three siblings shared a bedroom, the two boys in one bed and the two girls in the other, a setup which meant that they also shared their germs. Childhood infections were the norm—chicken pox, measles and whooping cough were three I remember him talking about. He and his older brother were clearly impacted by these experiences, with their deep dislike of germs and viruses eventually leading my uncle Irv to become a microbiologist and Sid, my father, to become an immunologist.

Since they grew up before the antibiotic era, they had only supportive measures like teas and compresses (and aspirin) to help them get through their infections. Their careers were devoted to using the tools of microbiology and immunology to find cures, but they died just as the psychological aspects of immunity were gaining recognition. We now know that stress impairs immune resilience, and we can acknowledge how important the terrain is. As Rudolf Virchow, an early proponent of the germ theory, stated in his later years, "If I could live my life over again, I would devote it to proving that germs seek their natural habitat—diseased tissues—rather than causing disease."

My own interest in mind/body medicine was nurtured by a psychiatry training program headed by the so-called father of psychoneuroimmunology (PNI), George F. Solomon, MD. So my work has been a continuation of my family's quest to conquer infectious diseases, but from the terrain rather than the germ point of view.

Unfortunately, America's response to Covid paid almost no attention to the terrain, the "host factors" that create a robust natural immunity, factors like nutrition, exercise, and stress management. Remember my med school microbiology professor who told us that his cold sores (oral herpes) would flare up every time his mother-in-law came to visit? Imagine if Dr. Fauci et al. had pursued that line of public health policy—self-management to enhance immune resilience.

Public health measures could have focused on such low-cost and readily scalable strategies as weight loss (obesity is a proven key risk factor), Vitamin D supplementation (low D levels are another risk factor for disease progression), exercise (a proven preventive factor), and stress management training—all of which are known to enhance immune resilience. Instead, the nation's strategy derived entirely from the germ theory model—vaccines, masking, and lockdowns, with a heavy overlay of fear to tie it all together. Physicians who played down the role of vaccines while instead stressing self-care were marginalized; some were even deplatformed, censored, or lost their licenses if they criticized the vaccines directly (see "The Disinformation Dozen" in chapter 2"). A good summary of the scientific and corporate issues driving the vaccine story can be found online (Kennedy 2022).

2. *Epidemics Spread by Fear*

Another form of epidemic is "emotional contagion"—the rapid spread of an emotion throughout a society by that culture's preferred means of communication, which could be anything from the grapevine of neighborhood porch-talk to the internet and social media. The emotion being spread can run the gamut from despair to joy, as seen with sports fans—with Covid, it's been fear. Of course, diseases and death will evoke fear as a matter of course, but at times the media's focus on Covid-19 related fears has seemed to be intentionally hyped, with daily case numbers and death counts having been featured on the front pages of our newspapers in full-color graphs and pie charts. It's ironic that a subject so scrutinized by science and medicine should still have so much uncertainty attached to the numbers we're barraged with, to the point that even seemingly unassailable stats like diagnosis and death rate have been questioned (i.e. "died from Covid" vs. "died with Covid"—was it a cause or an incidental finding?).

For context, Belgian psychologist Mattias Desmet warns us about the false sense of security our culture gets from relying too heavily on numbers and measurements and on the mechanical quantitative view of human beings (Desmet 2022). He talks of our "(im)measurable" universe, and gives examples of the statistical uncertainties that prevent us from gaining a clear picture of Covid: our overreliance on tests like the PCR which can be so sensitive as to detect no-longer-living viral fragments (thus generating misleadingly high rates of false-positive results), viewing a positive test result as an actual clinical case (even when no clinical symptoms of Covid are present), and so on. Daily media reports never made these distinctions clear, and thereby created a growing sense of fear and helplessness.

The negative health impact of lockdowns and isolation is significant, too, but has been mentioned only peripherally in the media. The collateral damage includes loneliness, food insecurity, depression, suicide, and substance abuse, all of which have increased dramatically in these Covid years and contribute an unknown but significant portion to the overall death rate. And the safety of the novel mRNA Covid vaccines (a credibility-enhancing misnomer, since true vaccines are attenuated forms of the whole virus, while this Covid shot contains only a fragment of their RNA blueprint) was not assessed in accordance with the usual vaccine pre-release standards that require years of testing on thousands of subjects. Instead, the pharmaceutical companies were granted an emergency use authorization (EUA) when the government decreed that there were no other available treatments for Covid (ignoring proven benefits of existing antiviral treatments, as well as ivermectin, hydroxychloroquine, and Vitamin D/C), thus providing a fast track for a pharmaceutical bonanza-via-monopoly and creating the vaccine-as-savior narrative in the American, and global, mindset. Imagine if only half of Moderna and Pfizer's $10 billion in vaccine profits had been directed to these preventive public health measures. . . .

We're left with what many have called an epidemic of fear, an emotional contagion that is accelerated by lack of clarity ("disinformation") and competing narratives. This uncertainty opens the door to some disturbing totalitarian trends in America today. Desmet has called this widely shared mindset a "mass formation"—a form of group hypnosis. And while the process of group trance can lead to states of shared euphoria (Leskowitz 2014b), the Covid media campaign resembles the opposite; tuning forks exposed to the frequency of fear will just as surely entrain as when they're exposed to euphoric vibrations. At times, the media campaign began to feel like the process used by totalitarian regimes to instill uncertainty and blame external forces (immigrants, Jews, viruses) in order to engender reliance on a powerful leader figure who steps forward as the savior.

3. Epidemics Spread by Dreamtime Intentions

So far in this section, we've looked at the possibility that groups of people in today's Covid era can be linked together by their common exposure to microscopic viruses and by a propaganda network that uses brainwashing techniques to inculcate widely held beliefs. But there's a third method of interconnection, a consciousness-based process that goes beyond the materialist model into the transpersonal world of mental radios.

Another possible explanation for epidemics was suggested forty years ago by an artist and poet living in upstate New York named Jane Roberts. She was a prolific author, with twenty-four titles to her credit and lifetime sales of over seven million books. But there was a twist—as she herself said, she didn't actually write her books. Rather, she transmitted information from a consciousness named Seth—"an energy personality essence no longer focused in physical matter"—who talked through her, literally taking over her voice in the process known as channeling (described in chapter 9). Apart from the fascinating questions that this process itself raises about "the nature of personal reality" (the title of one of her books), the explanation offered by Seth for mass events like epidemics is intriguing.

Seth endorses a multidimensional perspective in which our awareness isn't limited to our physical body and its sensory apparatus, but functions like the legendary mental radio. Not only does it receive input from a wide range of stations (frequencies) while ensconced in the person's skull, but it can also travel to other realms and dimensions. From Seth's perspective, consciousness is nonlocal, and space-time barriers do not exist. This point of view has been adopted by many professional organizations and international groups composed of thousands of PhDs, MDs, and clinicians and scholars researching these phenomena. Leading organizations include the following:

- International Association for Near-Death Studies
- Institute of Noetic Sciences
- Society for Scientific Exploration
- Manifesto for a Post-Materialist Science
- American Parapsychological Association
- The Monroe Institute
- Foundation for Shamanic Studies
- Multidisciplinary Association for Psychedelic Studies
- International Association for the Study of Dreams

Their findings are still routinely censored by Wikipedia and TED Talks and mocked by mainstream psychology, but a new consensus is emerging. Websites for these key professional organizations and journals are listed in Resources (p. 274), as the topic is clearly too vast to address here.

Seth describes how human beings interact with each other *en masse*—not at the pub, not online, but in the dream state! In this dreamworld space of shared consciousness—the astral plane—communication can occur between and among a wide range of people who have reached this gathering spot via "astral travel" (i.e., dreaming). Most participants may not even remember their nighttime journeys the next day, but group consensus can still be reached about many issues, including potential future events that are deemed by consensus to be worthy of manifesting in the physical reality of everyday life. In accordance with Seth's best-known teaching—"You create your own reality"—individuals participate in this group process to bring into existence the events of their individual lives, but also to create the reality of group events large and small, good and bad.

To be sure, some famous global epidemics had clear external, non-esoteric causative factors—the Spanish flu of 1918 blazed through a grieving, traumatized, malnourished, and immune-compromised world that had just endured the agony of its First World War. But the Sethian perspective on epidemics is startling in its near-total reversal of cause and effect, as will be seen in the following excerpts, all gathered from a series of Seth books that were published in the pre-Covid 1970s and '80s.

Fortunately, as we've seen throughout this book, the most powerful influence on the body's health is the status of its energy system, and the most powerful determinant of that flow is the alignment of beliefs and emotions. So it's not surprising that individual disease resistance is ultimately under our control, and, for Seth, our intention is the primary regulator of the process that harnesses the energies of the universe through our beliefs (conscious and unconscious), our emotions, and our vital energy to bring ideas into physical form. At one point, Seth summarizes this multidimensional creation process by saying that "The body is your living sculpture." In other words, the physical body is being continually shaped and reshaped, upgraded and degraded, by these inner processes, the same ones that underlie the creation of health, illness, and even epidemics. Here are some Sethian insights into the functioning of this process of group co-creation:

1. **The causes (of epidemics) are not biological**: "Biology is simply the carrier of a 'deadly intent' . . . No person becomes ill unless that illness serves a psychic or psychological reason" (Roberts 1972).

   This is also a foundational point in mind/body medicine: symptoms are sustained by the "secondary gain" or unconscious benefits that accrue to being in the patient role. Diseases arise as a biologic solution to psychospiritual challenges (Myss 1998).

2. **The environment in which an outbreak occurs invites the disorder**: Often outbreaks take place after ineffective political or social action, after some unified mass social protest has failed or is considered hopeless, or amid other economic/environmental stresses. They also often occur in wartime, targeting the population that has opposed the given conflict. This perspective aligns with the previous example of the post-World War I influenza epidemic. Mass hopelessness and despair were the triggers, the invitations to the microorganisms that allowed the virus a foothold into the body of humanity, the figurative and literal global terrain.

3. **Initially there is a psychic contagion**: "Despair moves faster than a mosquito, or any outward carrier of a given disease. The mental state brings about the activation of a virus that is, in those terms, passive . . ." (Roberts 1981).

   This is basically a metaphysical reframing of psychoneuroimmunology. The virus is no longer the "bad guy" but merely the mechanism for carrying out the unconscious inner intent.

4. **An epidemic is a warning**: "The epidemics then serve many purposes—warning that certain conditions will not be tolerated.

There is a biologic outrage that will be continually expressed until the conditions are changed . . . Your private mental states en masse bring about the mass cultural stance of your civilization . . . You live in a physical community, but you live first of all in community of thoughts and feelings. These trigger your physical actions. They directly affect the behavior of your body" (Roberts 1972).

Currently popular terms for this sort of intangible community are "interbeing" (Eisenstein 2013) and "interdependence." The added dimension from Seth's perspective is that these links are shared and experienced nonlocally, across space and time, in the transpersonal dimensions.

These are interesting ideas, if a bit far out—but how do they relate to Covid? Why would humanity want to generate such a devastating process, and why now? The answer can come on three levels: the political, the psychological and the metaphysical. The political/economic view sees current events as a power play in which the virus was strategically created by the "deep state" in a lab, as a bioweapon whose purpose is to create havoc and set the stage for world domination by this presumed global elite. As paranoid as this may sound, there are enough flaws in the official Covid narrative to make space for at least considering some parts of these alternate geopolitical possibilities, including the growing probability that the virus leaked from a Wuhan lab doing gain-of-function research on Covid (Berche 2023).

From the psychological perspective, we can view the world's many crises—from political to meteorological—as part of an emotional/energetic healing process. Our shadow side, the hidden parts of our collective psyche, have to emerge and be acknowledged before they can be healed and released. As with the "healing crisis" of homeopathy, worsening symptoms are paradoxically welcomed as a sign that the body's innate vitality is at work, clearing out residual toxic energy. And hopefully that's what this global reset of energy vibration is doing, clearing the way for "the more beautiful world our hearts know is possible" (Eisenstein 2013).

There's also the Sethian idea that humanity, at least in the dream state, is so deeply concerned about its current dire state of existence that a crisis had to be manifested as a call to action. Covid has certainly done so by raising many fundamental questions about how we operate as a civilization. It has given us a chance to reconsider our embrace of the disempowering medical model of disease causation, and our acceptance of the glaring inequalities in healthcare access, the growing chasm between the haves and the have-nots, and the sense that our "society" of people is being replaced by an "economy" of things that is destroying the Earth's ecosystem in the process. To the extent that we make the leap and become conscious of our true nature as interconnected multidimensional creators rather than isolated mechanical victims, Covid will have served its higher purpose.

THE AGE OF AQUARIUS

When the moon is in the Seventh House
And Jupiter aligns with Mars,
Then peace will guide the planets
And love will steer the stars.
THE AGE OF AQUARIUS, FROM *HAIR* (1967)

These memorable lyrics have inspired millions of people in the fifty-five years since they were first written, bringing the notion of a grand cultural transformation into the wider consciousness by presaging a coming New Age of love and peace. So I was disappointed to learn recently that, from an astrological point of view, the words are actually gibberish. The moon is in the seventh house (one of twelve such facets of the Zodiac) for several days each and every month as it orbits around the earth, and Jupiter aligns with Mars several times every decade because they're adjacent planets in the concentric rings of the solar system's planetary orbits. There's nothing at all unusual about this alignment, and it certainly doesn't portend a once-in-a-millennium type of event. But we can cut *Hair's* writers some slack here, as there's certainly room for a little poetic license in creating such an evocative image.

Regardless of the literal accuracy of the song, it makes sense to acknowledge the history and impact of astrology cross culturally as we consider the events and resources that push humanity forward on our journey. Astrology is one of many spiritual paths that sees mankind going through a series of steps or transitional eras on its way to full awakening, and has numerous forms and layers to its past and current expression:

- **Astrology**—The 2,100 year-long Age of Aquarius, marked by the values of collaboration and compassion, is in the process of emerging from the shadows of the Piscean Age of monotheism and hierarchy—regardless of how Jupiter and Mars align.
- **Astronomy**—New Age astronomers hold that the Earth is now entering a high-vibrational part of the galaxy called the Photon Belt, which it does every 26,000 years. This heightened energy will supposedly enable humans to transcend base emotions and lead lives of peace and love.
- **Judaism and Christianity**—Many branches of Orthodox Judaism and evangelical Christianity say that the Messiah of Biblical prophecy will be arriving soon, whether for the first time (Judaism) or in his return engagement (Christianity).
- **Hinduism**—This religious perspective holds that we are now ending the Kali Yuga (the Age of Darkness) and entering the Satya Yuga (the Age of Truth), with each of these ages lasting over 400,000 years.
- **Hopi**—These indigenous peoples have prophesied that a time of purification would come at the end of the Third World, or cycle of destruction, we are now immersed in.
- **Tibetan Buddhism**—Their texts have long said that "when the iron bird flies and horses run on wheels, the *dharma* (teachings)

will come to the land of the red men" (i.e., when airplanes and railroads come to North America). The spread of these teachings will lead to a long-awaited era of peace.

Perhaps it's not surprising that, along with these converging predictions from so many well-established spiritual traditions, we're also being inundated with channeled messages from such uncommon places as the Pleiades and Arcturus, advising us about energy activation and ascension into higher dimensions. In part because we're looking ever more desperately for help from ever more unlikely sources, interest in ETs is at an all-time high, and after decades of government suppression of UFO-related information, even NASA has acknowledged that there's more to the phenomenon than they've ever let on, with the summer of 2023 seeing the first in a series of open Congressional hearings and investigations. The New Age energies of the '60s—peace, love, understanding, and care of the Earth—have taken root, and either they will flourish or we will perish. So, whether we each consider ourselves to be different leaves on the same tree, different waves in the same ocean, different neurons in God's brain, or different raindrops falling from the same cloud, life energy will be the force that unites us as we carry out this vital healing mission together.

IN CONCLUSION

I'd like to finish by honoring all those who are on this collective journey to higher consciousness by taking some poetic license with the final stanza of Robert Frost's beloved poem, "The road not taken." I'll substitute we for the original I.

> *Two roads diverged in a wood, and we—*
> *We took the road less traveled by,*
> *And that has made all the difference.*

What started in the sixties as an uncoordinated upsurge in individual journeys of awakening has been transformed by our collective energetic bonding into a process of resonant global awakening. Our choice of the less-traveled road has made all the difference: Paradise is breaking through the old paradigm's paved-over parking lots, both conceptual and concrete—not just in figure 10.3, but the world over. Happy trails!

Figure 10.3—Paradise versus a parking lot

Acknowledgments

In the process of writing a book that covers so many phases of my life, I inevitably came to realize just how many people have had an important impact on my life. I'll acknowledge many of the key players here, with the understanding that many more who played a vital part had to be left on the cutting room floor (that's only a figure of speech).

To start with, I'm grateful for the ongoing love and support from my family, the nuclear one of mom Thelma, dad Sid, brother Andy and sister Laurie. And then later the extended family that included grandparents, cousins, aunts, uncles, and most importantly wife Doreen and children Shari and David, their partners Ethan and Emily and our companions Mango, Mirabelle and Tabha.

Education has played a key role in my life, with some key teachers deserving mention: my father, who introduced me to the world of science; my high school English teacher Hank Fortier for opening my eyes to Joseph Campbell and Carl Jung; Prof. Al Sorenson and his psychophysiology class in college; Earl Ettienne, my medical school mentor and the key way-shower into the world of holistic medicine. Several important teachers were nonphysical: Seth, Alexander, Abraham, Master Chi'ang, Monitor, and Mercator.

In the realm of energy medicine, Rev. Rosalyn Bruyere was my first and foremost teacher. Her community has been an ongoing support for over thirty years, especially Richard Allen, Mia Beale, Nancy

Needham, Ken Koles, and Reva Seybolt. I've been fortunate to have so many supportive energy-savvy friends and colleagues along the way, so special thanks go to Fella Cederbaum, Richard Allen and Lucie LeBlanc, Maureen Foye, Judy Tsafrir, Gloria Hemsher, the Spaulding Monday Evening Healers Group (Sheryl Lawrence, Martha Morgan, Eve Kennedy-Spaien, Jenn Jackson, Carol Seplowitz, and Janice Wesley), the Wednesday Night Supper Club (Tom Deters, Mel Glenn, Murdo Dowds, and Liz Selleck), the Saturday Morning Meditation Group, the Pain Team at Medford, the Integrative Medicine Task Force at Spaulding, the Sports Energy and Consciousness Group, and the ACEP Truth-Seekers.

Nuts and bolts technical help has come from Gail Fischer (photography), Shari Leskowitz (website), and Rosi Fatah (illustrations), while the team at Inner Traditions has been helpful every step of the way: Jon Graham, Erica Robinson, Courtney Jenkins, Ashley Kolesnik and especially my project editor Emilia Cataldo and copy editor Dorona Zierler. Thank you, thank you, thank you.

But in the end, I have to circle back to my primary group—my wife and kids. I've learned more about group energies, team chemistry, and the power of love from them than from any curriculum or teacher. I hope you guys have enjoyed the ride as much as I have!

Resources

ORGANIZATIONS

Princeton Engineering Anomalies Research (PEAR) Lab: pearlab.icrl.org

International Consciousness Research Lab (ICRL): icrl.org

Global Consciousness Project (GCP): teilhard.global-mind.org

HeartMath Institute and the Global Coherence Initiative (GCI): heartmath.org

Institute of Noetic Sciences (IONS): noetic.org

Association for Comprehensive Energy Psychology (ACEP): energypsych.org

Evolutionary Sports Collective (EVO Sports): evosportscollective.com

Consciousness and Healing Initiative (CHI): chi.is

Open Sciences: opensciences.org

The Crop Circle Connector: cropcircleconnector.com

Temporary Temples: temporarytemples.co.uk

JOURNALS

Explore: *The Journal of Science and Healing*: sciencedirect.com/journal/explore

Energy Psychology: Theory, Research and Treatment: energypsychologyjournal.org

International Journal of Healing and Caring (IJHC): ijhc.org
Journal of Scientific Exploration: Anomalistics and Frontier Science:
   journalofscientificexploration.org

MOVIES AND LECTURES

The Joy of Sox: Weird Science and the Power of Intention on
   TheJoyOfSoxMovie's YouTube channel.
Exposing Scientific Dogmas – Banned TED Talk – Rupert Sheldrake
   on After Skool's YouTube channel
*Frequency One with Dean Radin: Scientific Evidence for Psychic
   Phenomena?* on 11thstory's YouTube channel.
Thrive: What On Earth Will It Take?: freetothrive.com
Crop Circles: The Ultimate Undercover Investigation on Taysider's
   YouTube channel.

VIDEO DEMONSTRATIONS

The following video demonstrations can be found at
TheMysteryOfLifeEnergy.com/Resources

Perception via Phantom Limbs
Group Heart Coherence (from *The Joy of Sox*)
EFT for Athletes (from *The Joy of Sox*)
Measuring Fan Energy (from *The Joy of Sox*)
Starling Murmuration
Tapping (EFT) for Vets with PTSD
Aerial Survey of Crop Circles
Glass Harmonica—Mozart Adagio

Glossary of Acronyms

A&P: anatomy and physiology

ACEP: Association for Comprehensive Energy Psychology

ADD: Attention Deficit Disorder

AHHA: American Holistic Health Association

AI: Artificial Intelligence

AIT: Advanced Integrative Therapies

AMA: American Medical Association

APA: American Parapsychological Association

ASC: Altered States of Consciousness

ATP: adenosine triphosphate

CAM: Complementary and Alternative Medicine

CCDH: Center for Countering Digital Hate

CEO: Chief Executive Officer

CME: Continuing Medical Education

CST: CranioSacral Therapy

DSM: Diagnostic and Statistical Manual

ECT: electroconvulsive shock therapy

EEG: electroencephalogram

EFT: Emotional Freedom Techniques

EM: Energy Medicine

EMDR: Eye Movement Desensitization and Reprocessing

EMF: electromagnetic field

ESP: Extra Sensory Perception

ET: Extraterrestrials

EUA: emergency use authorization

FCC: Federal Communications Commission

FDA: Food and Drug Administration

FDR: Franklin Delano Roosevelt

fMRI: functional magnetic resonance imaging

FSS: Foundation for Shamanic Studies

G&S: Gilbert and Sullivan

GCI: Global Coherence Initiative

GCP: Global Consciousness Project

GRO: Greenwich Royal Observatory

HFA: home field advantage

HMI: HeartMath Institute

HPM: Human Potential Movement

HR: Human Resources

HRV: heart rate variability

IANDS: International Association for Near-Death Studies

IASD: International Association for the Study of Dreams

IM: Integrative Medicine

IMTF: Integrative Medicine Task Force

IONS: Institute of Noetic Sciences

JAMA: Journal of the American Medical Association

LEED: Leadership in Energy and Environmental Design

MAPS: Multi-disciplinary Association for Psychedelic Studies

MD: Medical Doctor

MIT: Massachusetts Institute of Technology

MIU: Maharishi International University

NASA: National Aeronautics and Space Administration

NDA: Nondisclosure Agreement

NGO: Nongovernmental Organization

NIH: National Institutes of Health

NCCIH: National Center for Complementary and Integrative Health

OAM: Office of Alternative Medicine

OT: Occupational Therapist

PBS: Public Broadcasting Station

PEAR: Princeton Engineering Anomalies Research

PEMS: pulsed electromagnetic stimulation

PLP: phantom limb pain

PM&R: Physical Medicine and Rehabilitation

PNI: psychoneuroimmunology

POV: Point of View

PT: Physical Therapist

PTSD: Post-Traumatic Stress Disorder

qEEG: quantitative electro-encephalogram

RN: Registered Nurse

RNA: ribonucleic acid

RNG: random number generator

SABR: Society for American Baseball Research

SAD: Seasonal Affective Disorder

SQUEBS: **S**ix **QU**ick **E**nergy **B**alancer**S**

SRH: Spaulding Rehabilitation Hospital

SSE: Society for Scientific Exploration

TCM: Traditional Chinese Medicine

TED: Technology, Entertainment, and Design (refers to recordings from talks recorded at the TED Conference)

TM: Transcendental Meditation

TPTB: The Powers That Be

TT: Therapeutic Touch

TUE: Therapeutic Use Exemption

UMMS: University of Massachusetts Medical School

UV: ultraviolet

VA: Veterans Administration

Bibliography

Abraham, Peter. 2022. "Do You Believe in Superstition? Alex Cora Shaved His Beard, and the Red Sox Bats Came Back to Life." *Boston Globe* website May 11, 2022. Accessed July 7, 2023.

Adams, Tim. 2022. "How Science Is Uncovering the Secrets of Stonehenge." *The Guardian* January 30, 2022. Accessed July 7, 2023.

Ahn, Andrew C., Agatha P. Colbert, Belinda J. Anderson, Ørjan G. Martinsen, Richard Hammerschlag, Steve Cina, Peter M. Wayne, and Helene M. Langevin. 2008. "Electrical Properties of Acupuncture Points and Meridians: A Systematic Review." *Bioelectromagnetics* 29, no. 4 (January 31): 245–56.

Al Azawe, L. 2020. "The Use of Psychic Perception to Enhance Athletic Performance: A Report from the Iraqi Olympic Committee." *International Journal of Healing and Caring* 21, no. 1.

Aleixo, Denise, Leoni V. Bonamin, Fabiana N. Ferraz, Franciele K. Da Veiga, and Silvana M. De Araújo. 2021. "Homeopathy in Parasitic Diseases." *International Journal of High Dilution Research* 13, no. 46 (November 30): 13–27.

Alexander, Sharon. 2020. "*Oh my God!*: Exploring Ecstatic Experience through the evocative technology of Gospel Choir." *International Journal of Healing and Caring* 20 no. 1 (January 9):1–15. Accessed July 12, 2023.

Amos, Tori. 2021. "Locked Down in Cornwall, the Sea Brought Me Energy and Spiritual Healing." *The Guardian* (November 1). Accessed July 7, 2023.

Baker, Robert. 1994. "End of the 'Crop Circle' Circus." *Skeptical Inquirer* 18 no. 2 (winter): 191–94.

Baranowski-Pinto, G., V. L. S. Profeta, Malcolm Newson, Harvey Whitehouse, and

Dimitris Xygalatas. 2022. "Being in a Crowd Bonds People via Physiological Synchrony." *Scientific Reports* 12, no. 1. Nature website January 12, 2022.

Barrett, Stephen. 2023. "Stephen Barrett, M.D. Biographical Sketch." Quackwatch website February 4, 2023. Accessed July 12, 2023.

Barnhart, Robert. 1995. *The Barnhart Concise Dictionary of Etymology: The Origins of American English Words.* New York: HarperCollins, 1995.

Beacham, Greg. 2020. "Rams Star Aaron Donald: Football without Fans 'Wouldn't Be Fun.'" *Boston Globe* May 22, 2020. Accessed July 12, 2023.

Beauregard, Mario, Gary E. Schwartz, Lisa A. Miller, Larry Dossey, Alexander Moreira-Almeida, Marilyn Schlitz, Rupert Sheldrake, and Charles T. Tart. 2014. "Manifesto for a Post-Materialist Science." *Explore: The Journal of Science and Healing* 10, no. 5 (September 1): 272–74.

Becker, Robert H. 1985. *Cross-Currents: The Promise of Electromedicine, the Perils of Electropollution.* UK: Penguin Press.

Belam, Martin, and Sean Ingle. 2022. "Not Just Snow: What's the Secret to Norway's Winter Olympic Success?" *The Guardian* February 18, 2022.

Bengston, W., and D. Krinsley. 2000. "The Effect of the 'Laying On of Hands' on Transplanted Breast Cancer in Mice." *Journal of Scientific Exploration* 14, no. 3 (January 1): 353–64.

Berche, Patrick. 2023. "Gain-of-Function and Origin of COVID-19." *La Presse Médicale* 52 (March 2023): 104167.

Bergland, Christopher. 2021. "'Runners High' Depends on Endocannabinoids (Not Endorphins)." *Psychology Today* (February 26).

Bergson, Henri. 1911. *Creative Evolution.* New York: Henry Holt and Co.

Beseme, Sarah, William F. Bengston, Dean Radin, Michael S. Turner, and John McMichael. "Transcriptional Changes in Cancer Cells Induced by Exposure to a Healing Method." *Dose-Response* 16, no. 3 (July 1, 2018): 1–8.

Beston, Henry. 1928. *The Outermost House: A Year of Life on the Great Beach of Cape Cod.* New York: Doubleday.

Bettelheim, Bruno. 1982. "Freud and the Soul." *The New Yorker,* February 22, 1982.

Bickert, Monika. 2021 "How we're taking action against vaccine misinformation superspreaders." Fb.com website August 18, 2021.

Bierle, Sarah. 2020. "The Foot That Is Gone Pains Me the Most." *Emerging Civil War* website, October 8, 2020.

Braden, Gregg. 1997. *Awakening to Zero Point: The Collective Initiation.* Radio Bookstore Press.

Braud, William, Donna Shafer, and Sperry Andrews 1993. "Reactions to an Unseen Gaze (Remote Attention): A Review, with New Data on Autonomic Staring Detection." *Journal of Parapsychology* 57: 373–90.

Broadhurst, Paul, and Hamish Miller. 1989. *The Sun and the Serpent,* UK: Pendragon Partnership; reprint MYTHOS.

Brooker, Charles. 1983. "Magnetism and the Standing Stones." *New Scientist* (January 13): 103.

Brown, Derren. 2011. "Trick of the Mind." Season 3, Episode 6, *Channel 4* YouTube video, August 17, 2011. Accessed July 12, 2023.

Brown, Toni A. 2011. "Phish, No Fear of Flying: An Interview with Mike Gordon" (Relix Revisited). *Relix Media* website, September 1, 2011.

Bulwer-Lytton, Edward, 1871. *The Coming Race.* London, UK: William Blackwood and Sons.

Burk, Larry. 2023." Transforming Symptoms of the Lower 4 Chakras in Physical Illness." (Substack). October 5, 2023.

Burke, John A., and Kaj Halberg. 2005. *Seed of Knowledge, Stone of Plenty: Understanding the Lost Technology of the Ancient Megalith-Builders.* San Francisco: Council Oak Books.

Burr, Harold Saxton. 1973. *The Fields of Life: Our Links with the Universe.* New York: Ballantine Books.

Burton-Hill, Clemency. 2022. "The Queen's Funeral Music Reveals Truths about Her—and All of Us." *Financial Times*, September 19, 2022. Accessed July 12, 2023.

Byrd, Randolph C. 1988. "Positive Therapeutic Effects of Intercessory Prayer in a Coronary Care Unit Population." *Southern Medical Journal* 81, no. 7 (July 1): 826–29.

Carayol, Tumaini. 2023. "Cameron Norrie Calls for Video Replays after Row in Lucas Pouille Win." *The Guardian*, May 31, 2023.

Carey, Benedict. 2004. "Researcher Pulls His Name from Paper on Prayer and Fertility." *New York Times,* December 4, 2004.

Callahan, Maureen. 2021. "Anti-Vax, Conspiracy Theorist, Collector of Dead Animals—RFK Jr. Is the Dumbest Kennedy." *New York Post*, September 24, 2021.

Castaneda, Carlos. 1974. *Tales of Power.* New York: Simon and Schuster.

Chamberlin, Kent. 2021. *Electromagnetic Radiation and Health.* "NH Commission Setback Justification." YouTube video. December 28, 2021.

Champoux, Peter. 1999. *Gaia Matrix: Arkhom and the Geometries of Destiny in the North American Continent.* Washington, Mass.: Franklin Media.

Chapman, Edward H., Richard J. Weintraub, Michael A. Milburn, Therese O'Neil Pirozzi, and Elaine Woo. 1999. "Homeopathic Treatment of Mild Traumatic Brain Injury: A Randomized, Double-Blind, Placebo-Controlled Clinical Trial." *Journal of Head Trauma Rehabilitation* 14, no. 6 (December 1): 521–42.

Church, Dawson. 2009. "The Effect of EFT (Emotional Freedom Techniques) on Athletic Performance: A Randomized Controlled Blind Trial." *The Open Sports Sciences Journal* 2, no. 1 (January 4): 94–99.

Church, Dawson, Crystal Hawk, Audrey J. Brooks, Olli T. Toukolehto, Maria Wren, Ingrid Dinter, and Phyllis K. Stein. 2013. "Psychological Trauma Symptom Improvement in Veterans Using Emotional Freedom Techniques." *Journal of Nervous and Mental Disease* 201, no. 2 (February 1): 153–60.

Church, Dawson. 2014. *The Genie in Your Genes: Epigenetic Medicine and the New Biology of Intention* 3rd ed. Santa Rosa, Calif.: Energy Psychology Press.

Clow, Barbara Hand. 1995. *The Pleiadian Agenda: A New Cosmology for the Age of Light.* Rochester, Vt.: Bear & Company.

Cohen, Kenneth S. 1999. *The Way of Qigong: The Art and Science of Chinese Energy Healing.* New York: Wellspring/Ballantine.

Colter, Wendie, and Paul Mills. 2021. "Assessing the Accuracy of Medical Intuition: A Subjective and Exploratory Study." *Journal of Alternative and Complementary Medicine* 26, no. 12 (December 9): 1130–35.

Conan Doyle, Arthur. 1926/1995. "The Land of Mist," in *The Lost World and Other Stories*, 368. UK: Wordsworth Editions Limited.

Conan Doyle, Arthur. 1890/1975. *The Sign of Four*, 173. Berkley Medallion Books.

Conrad, Marissa. 2021. "Unmute that Zoom Call." *Boston Globe*, April 21, 2021. Accessed July 12, 2023.

Coodley, Lauren. 2013. *Upton Sinclair: California Socialist, Celebrity Intellectual*, 104–06. Lincoln, Neb.: Bison Books.

Cowen, Leslie. 2007. "The Crop Circle Mystery." *Gazette and Herald*, July 20, 2007.

Crompton, Sarah. 2023. "Alessandra Ferri: Dancing with Nureyev, I Swear He Had an Energetic Aura around Him." *The Guardian*, February 26, 2023.

Csikszentmihalyi, Michael. 2014. *Flow: The Psychology of Optimal Experience.* New York: Harper Perennial.

de Chardin, Teilhard. 1955. *The Phenomenon of Man*. New York: Harper Collins.

de Kruif, Paul. 1926. *Microbe Hunters*. New York: Blue Ribbon Books.

Delaney, Brigid. 2022. "Moshpits, Megacities and Mecca: 'Overwhelming' New Film Captures the Brutality and Beauty of Crowds." *The Guardian* website, August 5, 2022.

Delbanco Tom, R. Ivker, A. Relman, and Eric Leskowitz. 2000. "Complementary and Alternative Therapies and the Question of Evidence." *Advances in Mind/Body Medicine,* 16:244–60.

Delorme, Arnaud, Cassandra Vieten, Dean Radin, and Loren Carpenter. 2017. "Special Report: Collective Consciousness at Burning Man." *Institute of Noetic Sciences,* March 2, 2017. Accessed July 12, 2023.

DeMeo, James. 2010. *The Orgone Accumulator Handbook*. Natural Energy Works.

Desmet, Mattias. 2022. *The Psychology of Totalitarianism*. White River Junction, Vt.: Chelsea Green Publishing.

Diamond, John. 1983. *Life Energy: Using the Meridians to Unlock the Hidden Power of Your Emotions*. St. Paul, Minn.: Paragon House Publishers.

Diamond, John. 2021. "Dr. John Diamond | Healing from Within," May 12. Accessed July 12, 2023. Dr. John Diamond website.

Dies, Albert C. 1810. *Biographical Accounts of Franz Haydn*. Vienna, Austria: Camesinaische Buchhandlung.

Djaali, Wahunihnsih, Bazzar Ari Migrah, Fatah Nurdin, Yasep Setiakarnawijaya. 2023. "Press-Tack Needle Acupuncture Reduces Postexercise Blood Lactic-Acid Levels in Sports Students." *Medical Acupuncture,* 35:5, October 17, 2023.

Doidge, Norman. 2007. *The Brain that Changes Itself: Stories of Personal Triumph from the Frontiers of Brain Science*. New York: Penguin Books.

Donaldson, I. M. L. 2005. "Mesmer's 1780 Proposal for a Controlled Trial to Test His Method of Treatment Using 'Animal Magnetism.'" *Journal of the Royal Society of Medicine* 98, no. 12 (December 1): 572–75.

Dorland, Jason. 2018. "Love as a Competitive Strategy." Your Mindset website, July 22, 2018.

Dossey, Larry. 2014. *One Mind: How Our Individual Mind Is Part of a Great Consciousness, and Why It Matters*. New York: Hay House.

Eden, Donna, 2021. "Personal communication." Interview by Eric Leskowitz, March 9, 2021.

Eden, Donna. n.d. "Eden Method—Empowering You with Energy Tools for a Life You Desire." Eden Method website.

Eden, Donna, and David Feinstein. 1998. *Energy Medicine: Balancing Your Body's Energies for Optimal Health, Joy and Vitality*. New York: Penguin Books.

Edwards, Stephen D. 2020. "Global Coherence, Healing Meditations Using HeartMath Applications during COVID-19 Lockdown." *Theological Studies/Teologiese Studies* 76, no. 1 (December 17).

Ehrenreich, Barbara. 2006. *Dancing in the Streets: A History of Collective Joy*. New York: Holt Paperback.

Eisenstein, Charles. 2013. *The More Beautiful World Our Hearts Know Is Possible*. Berkeley, Calif.: North Atlantic Books.

Ellenberger, Henri. 1970. *The Discovery of the Unconscious: The History and Evolution of Dynamic Psychiatry*. New York: Basic Books.

Feinstein, David. 2022. "The Energy of Energy Psychology." *OBM Integrative and Complementary Medicine* 7, no. 2 (January 2): 015.

Feinstein, David L. 2022. "Uses of Energy Psychology Following Catastrophic Events." *Frontiers in Psychology* 13 (April 25).

Feinstein, David. 2024. *Tapping: Harness the Transformative Power of Energy Psychology*. Louisville, Colo.: Sounds True.

Field, Tiffany. 2010. "Preterm Infant Massage Therapy Research." *Infant Behavioral Development* 33, no. 2: 115–24.

Flemming, Alexandra. 2020. "The Origins of Vaccination." *Nature Research*, September 28, 2020.

Flor, Herta, Lone Nikolajsen, and Troels S. Jensen. 2006. "Phantom Limb Pain: A Case of Maladaptive CNS Plasticity?" *Nature Reviews Neuroscience* 7, no. 11 (November 1): 873–81.

Ford, Scott. 2014. *Welcome to the Zone: Peak Performance Redefined*. Outskirts Press.

Ford, Scott. 2022. Personal communication. Interview by Eric Leskowitz, May 9.

Fortier, Hank. Personal communication. Interview by Eric Leskowitz, 1967.

Fortune, Dion. 1936. *The Goat-Foot God*. London: Williams and Norgate.

Frantzis, Bruce. 2019. "Opening the Energy Gates of Your Body Qigong." Energy Arts website, December 2019.

Freud, Anna. 1939/1992. *The Ego and the Mechanisms of Defense*. Milton Park, UK: Routledge.

Galbraith, Robert. 2020. *Troubled Blood*, 111. London: Sphere Books.

Gale, Richard. 2019. "Stephen Barrett: Wikipedia's Agent Provacateur against Natural Medicine." Skeptical about Skeptics website, April 1, 2019.

Gallwey, Timothy. 1972. *The Inner Game of Tennis*. New York: Random House.

Gantt, Darin. 2020. "Robbie Gould: Lack of Fans Will Make the Wind Different in Stadiums—NBC Sports." *NBC Sports* website, August 19, 2020.

Gardner, Martin. 1957. *Fads and Fallacies in the Name of Science*. New York: Courier Dover Editions.

Gauld, Alan. 1992. *A History of the Hypnotism*, 207. Cambridge, UK: Cambridge University Press.

Geddes, Martin. 2021. "Censored! Amazon's River of Red Ink." Martin Geddes website, April 30, 2021.

Gerber, Richard. 1996. *Vibrational Medicine: New Choices for Healing Ourselves*. Rochester, Vt.: Bear & Company.

Glickman, Michael, 2009. *Crop Circles: The Bones of God*. UK: Frog Books.

Goldman, Jonathan. Personal communication, July 14, 2023.

Goldman, Jonathan and Andi Goldman. 2017. *The Humming Effect: Sound Healing for Health and Happiness*. Rochester, Vt.: Healing Arts Press.

Goleman, Daniel, and Richard Davidson. 2017. *Altered Traits: Science Reveals How Meditation Changes Your Mind, Brain, and Body*. New York: Avery/Penguin Random House.

Grad, Bernard. 1965. "Some Biological Effects of the 'Laying on of Hands': A Review of Experiments with Animals and Plants," *Journal of the American Society for Psychical Research*, 59.

Grady, Harvey, and Julie Grady. 2009. *Explore with Monitor: Book 1: Lessons for Freeing Yourself*. iUniverse.

Graham, Bryan Armen. 2020. "The Most Notable US Athletes of 2020: No 2—Mariah Bell, the Edge of Glory." *The Guardian* website, December 30, 2020.

GreenMedInfo, 2023. "Children's Defense Fund Lawsuit Against Legacy Media for Collective Censorship of Online News; Anti-Trust Collaboration and Suppression." February 15, 2023.

Grattan, C. H. 2023. "Albert Einstein Endorsed a Popular Psychic in 1932. This Is the Controversy that Ensued." *The New Republic*, July 14, 2023.

Grafton, Sue. 1985. *B is for Burglar*. New York: Holt, Rhinehart, and Winston.

Greene, Debra. 2009. *Endless Energy: The Essential Guide to Energy Health*. Maui, Hi.: Metacomm Media.

Grey, Alex. 1990. *Sacred Mirrors: The Visionary Art of Alex Grey*: Rochester, Vt.: Inner Traditions.

Grey, Alex. Personal Communication, Interview by Eric Leskowitz, 2008.

Gronowicz, Gloria, Ankur Jhaveri, Libbe W. Clarke, Michael S. Aronow, and Theresa H. Smith. 2008. "Therapeutic Touch Stimulates the Proliferation of Human Cells in Culture." *Journal of Alternative and Complementary Medicine* 14, no. 3 (April 1): 233–39.

Grouios, George, Klio Semoglou, Katerina Mousikou, Konstantinos Chatzinikolaou, and Christos Kabitsis. 1997. "The Effect of a Simulated Mental Practice Technique on Free Throw Shooting Accuracy of Highly Skilled Basketball Players." *Journal of Human Movement Studies* 33(3): 119–38.

Gruder, David. 2014. TEDx Talks. "The Hijack: David Gruder at TEDxEncinitas." YouTube video, April 11, 2014. Accessed July 12, 2023.

Guardian Sport, 2020. "Mark Allen and Shaun Murphy Crash out of World Championship on Day of Shocks." *The Guardian*, August 5, 2020.

Hagelin, John S., Maxwell Rainforth, Kenneth L. Cavanaugh, Charles N. Alexander, Susan F. Shatkin, John Davies, Anne O. Hughes, Emanuel R. Ross, and David W. Orme-Johnson. 1999. "Effects of Group Practice of the Transcendental Meditation Program on Preventing Violent Crime in Washington, D.C.: Results of the National Demonstration Project, June–July 1993." *Social Indicators Research* 47, no. 2 (January 1): 153–201.

Hahn, Julie M., Patricia A. Reilly, and Teresa K. Buchanan. 2014. "Development of a Hospital Reiki Training Program." *Dimensions of Critical Care Nursing* 33, no. 1 (January 1): 15–21.

Hall, Howard, N. S. Don, J. N. Hussein, E. White, and Robert Hostoffer. 2001. "The Scientific Study of Unusual Rapid Wound Healing: A Case Report." *Advances in Mind/Body Medicine*, 17(3): 203–9.

Hammerschlag, Richard, Benjamin L. Marx, and Mikel Aickin. 2014. "Nontouch Biofield Therapy: A Systematic Review of Human Randomized Controlled Trials Reporting Use of Only Nonphysical Contact Treatment." *Journal of Alternative and Complementary Medicine* 20, no. 12 (December 1): 881–92.

Hammerschlag, Richard, Michael Levin, Rollin McCraty, Namuun Bat, John A. Ives, Susan K. Lutgendorf, and James L. Oschman. 2015. "Biofield Physiology: A Framework for an Emerging Discipline." *Global Advances in Health and Medicine* 4, no. 1 suppl. (January 1).

Hampton, Kevin C. 2012. "When the Yips Afflict Athletes, They Call Greg Warburton for Help." *Corvallis Gazette-Times*, July 29, 2012.

Harvard Health Publishing, 2021. "The power of the placebo effect." December 13, 2021.

Hawkins, Ed. 2019. *The Men on Magic Carpets: Searching for the Superhuman Sports Star*. London, UK: Bloomsbury Sport.

Hein, Simeon, and Russell Ron. 2002. "Electromagnetic and Crystalline Properties of Crop Formations." *The Institute for Resonance*, April 20, 2002.

Heinrich, Amelie, Florian H. Müller, Oliver Stoll, and Rouwen Cañal-Bruland. 2021. "Selection Bias in Social Facilitation Theory? Audience Effects on Elite Biathletes' Performance Are Gender-Specific." *Psychology of Sport and Exercise* 55 (July 1): 101943.

Herrigel, Eugen. 1953. *Zen in the Art of Archery*. New York: Pantheon Books.

Himmelsbach, Adam. 2020. "As We Consider a Future without Fans in the Stands, What Does That Mean for Athletes?" *Boston Globe*, May 24, 2020.

Hinson Glenn. 2000. *Fire in My Bones: Transcendence and the Holy Spirit in African American Gospel*. Philadelphia: University of Pennsylvania Press.

Hubacher, John. 2015. "The Phantom Leaf Effect: A Replication (Part 1)," *Journal of Alternative and Complementary Medicine*, 21(2): 83–90.

Huxley, Aldous. 1954. *The Doors of Perception*. UK: Chatto and Windus.

Ingle, Sean. 2022. "Defiant Sebastian Coe Flies Flag for London Olympics with a Murky Legacy." *The Guardian*, July 26, 2022.

Ireland, M. 1998. "An Even Closer Look at Therapeutic Touch." *Journal of the American Medical Association* 280(22): 1905–8.

Isaacs, J., and Terry Patten. 1991. "A Double Blind Study of the 'Biocircuit,' A Putative, Subtle-Energy-Based Relaxation Device." *Subtle Energies and Energy Medicine*, 2(2).

Jain, Shamini. 2021. *Healing Ourselves: Biofield Science and the Future of Health*. Sounds True.

Jain, Shamini, John A. Ives, Wayne B. Jonas, Richard Hammerschlag, David J. Muehsam, Cassandra Vieten, Daniel Vicario, Deepak Chopra, Rauni Pritten King, and Erminia M. Guarneri. 2015. "Biofield Science and

Healing: An Emerging Frontier in Medicine." *Global Advances in Health and Medicine* 4, no. 1 suppl. (January 1).

Jensen, Maureen, Michael Brant-Zawadzki, Nancy Obuchowski, Michael T. Modic, Dennis Malkasian, and Jeffrey S. Ross. 1994. "Magnetic Resonance Imaging of the Lumbar Spine in People without Back Pain." *New England Journal of Medicine*, 331: 69–73.

Ji, Sayer. 2021. "BREAKING: 'Disinformation Dozen': A 'Faulty Narrative' with No Evidence, Says Facebook, Despite 16,000 News Headlines." *Green Med Info*, August 19, 2021.

Joy, James. 2016. *The Quantum Sculler*. James Jon/VanAmringe, 2nd ed.

Judith, Anodea. 1997. *Eastern Body, Western Mind: Psychology and the Chakra System as a Path to the Self.* Berkeley, Calif.: Ten Speed Press.

Judith, Anodea. 2018. *Charge and the Energy Body: The Vital Key to Healing Your Life, Your Chakras and Your Relationships.* New York: Hay House.

Jung, Carl. 1969. *Man and His Symbols.* New York: Doubleday.

Kantor, Jerry. 2022. *Sane Asylums: The Success of Homeopathy before Psychiatry Lost Its Mind.* Rochester, Vt.: Healing Arts Press.

Kaplan, Fred. 1975. *Dickens and Mesmerism: The Hidden Springs of Fiction.* Princeton, N.J.: Princeton University Press.

Kaptchuk, T. 1983. *The Web that Has No Weaver: Understanding Chinese Medicine.* New York: McGraw Hill.

Kennedy, Robert F., Jr. 2021. *The Real Anthony Fauci: Bill Gates, Big Pharma, and the Global War on Democracy and Public Health.* Washington, D.C.: SkyHorse Publishing.

Kennedy, Robert F., Jr. 2022. *A Letter to Liberals: Censorship and COVID: An Attack on Science and American Ideals.* Washington, D.C.: Children's Health Defense.

King, Stephen. 2008. *Duma Key.* Pocket Books, Lancaster, Penn..

Kisner, Jordan. 2020. "Reiki Can't Possibly Work, So Why Does It?" *The Atlantic,* April, 2020. *The Atlantic* website.

Klagsbrun, Joan. 2001. "Listening and Focusing." *Nursing Clinics of North America* 36, no. 1 (March 1): 115–29.

Koestler, Arthur. 1967. *The Ghost in the Machine.* New York: Macmillan.

Kolata, Gina. 1998. "A Child's Paper Poses a Medical Challenge." *The New York Times*, April 1, 1998. *New York Times* website.

Kong, Jian, Ted J. Kaptchuk, Julia Webb, Jiang-Ti Kong, Yuka Sasaki,

Ginger Polich, Mark Vangel, Kenneth K. Kwong, Bruce R. Rosen, and Randy L. Gollub. 2009. "Functional Neuroanatomical Investigation of Vision-Related Acupuncture Point Specificity—A Multisession FMRI Study." *Human Brain Mapping* 30, no. 1 (January 1): 38–46.

Kotler Steven. 2014. *The Rise of Superman: Decoding the Science of Ultimate Human Performance.* London, UK: Quercus Publishing.

Krech, David. 1968. "Assault on the Citadel." (Book Reviews: *The Ghost in the Machine).*" *Science* 160, no. 3828: 649–50.

Krieger, Dolores. 1979. *The Therapeutic Touch: How to Use Your Hands to Help or to Heal.* New York: Touchstone.

Krippner, Stan, and Rubin D. (eds.). 1974. *The Kirlian Aura: Photographing the Galaxies of Life.* New York: Doubleday.

Laing, Olivia. 2021. "Wilhelm Reich: The Strange, Prescient Sexologist Who Sought to Set Us Free." *The Guardian*, April 17, 2021.

Lane, Anthony. 2023. "Zonked: The Exhausting History of Fatigue." *The New Yorker*, April 17, 2023: 56.

Langevin, Helene, David L. Churchill, Junru Wu, Gary J. Badger, Jason A. Yandow, James R. Fox, and Martin H. Krag. 2002. "Evidence of Connective Tissue Involvement in Acupuncture." *The FASEB Journal* 16, no. 8 (April 10): 872–74.

Laskow, Leonard. 1992. *Healing with Love: A Breakthrough Mind/Body Medical Program for Healing Yourself and Others.* Bloomington, Ind.: Author's Choice Press.

Lazar, Sara W., Catherine E. Kerr, Rachel M. Wasserman, Jeremy Gray, Douglas N. Greve, Michael T. Treadway, Metta McGarvey, et al. 2005. "Meditation Experience Is Associated with Increased Cortical Thickness." *Neuroreport* 16, no. 17 (November 28): 1893–97.

Leigh, Jill. 2021. "Keeping It Clean in Your Practice with Energy Hygiene." Energy Healing Institute website.

Leroy, Axelle, and Guy Cheron. 2020. "EEG Dynamics and Neural Generators of Psychological Flow during One Tightrope Performance." *Scientific Reports* 10, no. 1 (July 24).

Leskowitz, E. 1997. "Metaphors in the Teaching of Holistic Medicine." *Alternative Therapies in Health and Medicine,* 1 Jul 1997, 3(4): 111, 112.

Leskowitz, E. 1998. "Un-Debunking Therapeutic Touch." *Alternative Therapies in Health and Medicine* 4, no. 4:101–2.

Leskowitz, E. 2021. "Altered States of Consciousness: Chess in the Zone." *Explore: The Journal of Science and Healing* 18, no. 4 (July–August 2022): 395–98.

Leskowitz, E. 1999. "Life Energy for the Hemispherically Challenged." Keynote address, AHMA, Seattle.

Leskowitz, E. 2008. "Group Coherence and Target Subject Physiology." *Subtle Energies and Energy Medicine,* 18(3): 1–12.

Leskowitz, E. 2014a. "The 2013 World Series of Baseball: A Trojan Horse for Consciousness Studies." *Explore: The Journal of Science and Healing* 10, no. 2 (March–April 2014): 125–27.

Leskowitz, E. 2014b. "The Energetics of Group Trance: New Research, Applications, and Implications." *Energy Psychology Journal,* 6(2): 34–43.

Leskowitz, E. 2014c. "Harvard Doc Says Wikipedia Unfair to Alternative Therapies." *Commonhealth,* November 28, 2014. WBUR website.

Leskowitz, E. 2014d. "Phantom Limb Pain: An Energy/Trauma Model." *Explore: Journal of Science and Healing* 10, no. 6 (November–December 2014): 389–97.

Leskowitz, Eric. 2017. "The Zone: A Measurable (and Contagious) Exemplar of Mind/Body Coordination." *Journal of Alternative and Complementary Medicine* 23, no. 5 (May): 324–5.

Leskowitz, E. 2019. "The History of Integrative Medicine at Spaulding Rehabilitation Hospital." Spaulding Rehabilitation Network website, April.

Leskowitz, E. 2019. "Mesmer Reconsidered: From Animal Magnetism to the Biofield." *Explore: The Journal of Science and Healing,* 15(2): 95–97.

Leskowitz, E. 2022. "A Cartography of Energy Medicine: From Subtle Anatomy to Energy Physiology." *Explore: The Journal of Science and Healing* 18, no. 2 (March–April 2022): 152–64.

Leskowitz, E. 2023. "EFT in Hollywood." *ACEP News,* April, 2023. Energypsych.org.

Leskowitz, E, Foye M., Maliszewski M. 2016. "A Pilot Study Evaluating the Efficacy of a Biofield Energy Therapy in Chronic Pain Patients." *American Journal of Physical Medicine & Rehabilitation / Association of Academic Physiatrists* 93, no. 3 suppl. (March): a70.

Lettvin, J. Y., Humberto R. Maturana, Warren S. McCulloch, and William M. Pitts. 1959. "What the Frog's Eye Tells the Frog's Brain." *Proceedings of the IRE* 47, no. 11 (November 1): 1940–51.

Lewis, Sinclair. 1925. *Arrowsmith*. New York: Harcourt Brace & Co.

Levengood, William. 1994. "Anatomical Anomalies in Crop Formation Plants," *Physiologica Plantarum* 92, no. 2: 356–63.

Levin, Michael. 2014. "Molecular Bioelectricity: How Endogenous Voltage Potentials Control Cell Behavior and Instruct Pattern Regulation in Vivo." *Molecular Biology of the Cell* 25, no. 24 (December): 3835–50.

Levine, Peter. 1997. *Waking the Tiger: Healing Trauma*. Berkeley, Calif.: North Atlantic Books.

Liew, Jonathan. 2020. "Football Is Back and We Are Grateful but a Crowd Is Not a Sound Effect." *The Guardian*, June 30, 2020.

Litscher, Gerhard. 2005. "Infrared Thermography Fails to Visualize Stimulation-Induced Meridian-like Structures." *BioMedical Engineering OnLine* 4: 38.

Liu, Siyuan, Ho Ming Chow, Yisheng Xu, Michael G. Erkkinen, Katherine E. Swett, Michael W. Eagle, Daniel Rizik-Baer, and Allen R. Braun. 2012. "Neural Correlates of Lyrical Improvisation: An FMRI Study of Freestyle Rap." *Scientific Reports* 2, no. 1 (November 15).

Llewellyn-Edwards, T., and M. Llewellyn-Edwards. 2012. "The Effect of EFT (Emotional Freedom Techniques) on Soccer Performance." *Fidelity: Journal for the National Council of Psychotherapy* 47, 14–19.

Lonegren, Sig. 1996. *Spiritual Dowsing: Tools for Exploring the Intangible Realms*. Glastonbury, UK: Gothic Images Publications.

Lovelock, James. 1976. *Gaia: A New Look at Life on Earth*. Oxford, UK: Oxford University Press.

Loza, Boris. 2019. "The Hidden Power of Ancient Megaliths." *Nexus* 26, no. 3 (April–May): 59–63.

Lucia, René. 2009. *Unplugging the Patriarchy,* Crown Chakra Publishing.

Luckett, Richard. 1995. *Handel's Messiah: A Celebration*. Boston, Mass.: Mariner Books.

Maeterlink, Maurice. 1913. *The Life of the Bee*. New York: Dodd, Mead, and Company.

Marciak-Kozlowska, Janina, and Miroslaw Kozlowski. 2018. "On the Emergence of the Crop Circle." *Journal of Consciousness Exploration and Research* 9, no. 8: 800–805.

Mavor, James W. and Byron E. Dix. 1989. *Manitou: The Sacred Landscape of New England's Native Civilization*. Rochester, Vt.: Inner Traditions.

McCraty, Rollin. 2015. *The Science of the Heart: An Overview of the Role of the Heart in Human Performance.* HeartMath website, November 1, 2015.

McCraty, Rollin. 2017. "New Frontiers in Heart Rate Variability and Social Coherence Research: Techniques, Technologies, and Implications for Improving Group Dynamics and Outcomes." *Frontiers in Public Health* 5 (October 12).

McCraty, Rollin, Michael M. Atkinson, Inga Timofejeva, Roza Joffe, Alfonsas Vainoras, Mantas Landauskas, Abdullah A. Al-Abdulgader, and Minvydas Ragulskis. 2018. "The Influence of Heart Coherence on Synchronization between Human Heart Rate Variability and Geomagnetic Activity." *Journal of Complexity in Health Sciences* 1, no. 2 (December 31): 42–48.

McCurdy, Garvin. 2007. "The Inductive Chain Model. Harold Burr, Subtle Energies and the Insufficiency Theorem." *Subtle Energies and Energy Medicine* 18, no. 2: 63–73.

McCusick, Elaine. 2021. *Tuning the Human Biofield: Healing with Vibrational Sound Therapy*, 2nd ed. Rochester, Vt.: Healing Arts Press.

McLandsborough, L. 2007. "Why Do Apple Slices Turn Brown after Being Cut?" *Scientific American* website, July 30, 2007.

McTaggart, Lynne. 2008. *The Intention Experiment: Using Your Thoughts to Change Your Life and Your World.* New York: Atria Books.

McTaggart, Lynne. 2011. "The Experiments." Lynne McTaggart website, February 2, 2011.

Meehan, T. 1998. "Therapeutic Touch as a Nursing Intervention." *Journal of Advanced Nursing*, 28(1):117–25.

Meggyesy, David. 2014. "Concussion and the Zone," in Leskowitz, E. (Ed.) *Sports, Energy and Consciousness: Awakening Human Potential through Sport.* North Charleston, S.C.: CreateSpace Publishing.

Melville, Herman. 1851. *Moby-Dick; or The Whale.* London, UK: Richard Bentley. Chap. 108.

Melzack, Ronald, and Patrick D. Wall. 1996. "Pain Mechanisms: A New Theory." Pain Forum 5, no. 1 (March 1): 3–11.

Mercola, J. 2020. "Spy Agencies Threaten to Take Out Mercola." *GreenMedInfo*, December 7.

Michell, John (Ed.). 1991. *Dowsing the Crop Circles: New Insights into the Greatest of Modern Mysteries.* Glastonbury, UK: Gothic Image Publications.

Miller, Anna M. 2018. "Does Your Child Need an Energy Healer?" *US News and World Report*, June 5, 2018.

Millman, Dan. 1980. *The Way of the Peaceful Warrior*. Los Angeles: J. Tarcher.

Moga, Margaret. 2022. "Is There Scientific Evidence for Chakras?" *International Journal of Healing and Caring* 22, no. 2: 39–45.

Monitor, 2018. *Explore!* 18, no. 9: 12.

Muir, John. 1911/2004. *My First Summer in the Sierra*. Mineola, NY: Dover Books.

Mumford, George. 2015. *The Mindful Athlete: The Secret to Pure Performance*. Berkeley: Parallax Press.

Murphy, Michael. 1971. *Golf in the Kingdom*. New York: Penguin Books.

Murphy, Michael, and Rhea A. White. 2011. *In the Zone: Transcendent Experience in Sports*. New York: Open Road Media.

Myers, Benjamin. 2022. "Anonymous, Anti-Capitalist and Awe-Inspiring: Were Crop Circles Actually Great Art?" *The Guardian*, May 5.

Myss, Caroline. 1998. *Why People Don't Heal and How They Can*. Easton, Penn.: Harmony Press.

Myss, Caroline M., and C. Norman Shealy. 1988. *The Creation of Health: The Emotional, Psychological, and Spiritual Responses That Promote Health and Healing*. Walpole, N.H.: Stillpoint Publishing.

Naeser, Margaret, 2006. "Photobiomodulation of Pain in Carpal Tunnel Syndrome: Review of Seven Laser Therapy Studies." *Photomedicine and Laser Surgery* 24, no. 2 (April): 101–10.

Nakashima, Hiroaki. Yukawa, Yasutsugu. Suda, Kota, Masatsune Yamagata, Takayoshi Ueta, and Fumiko Kato. 2015. "Abnormal Findings on Magnetic Resonance Images of the Cervical Spines in 1211 Asymptomatic Subjects," *Spine* 15, 40(6): 392–8.

Nelson, Roger. 2017. "Princeton Engineering Anomalies Research," *PSI Encyclopedia* website, London: The Society for Psychical Research October 31, last updated July 4, 2023.

Nelson, Roger. 2019. *Connected: The Emergence of Global Consciousness*. Princeton, N.J.: ICRL Press.

Nestor, James. 2020. *Breath: The New Science of a Lost Art*. New York: Riverhead Books.

Newman, Hugh. 2017. *Stone Circles*. Glastonbury, UK: Wooden Books.

Norsworthy, Cameron, Ben Jackson, and James A. Dimmock. 2021.

"Advancing Our Understanding of Psychological Flow: A Scoping Review of Conceptualizations, Measurements, and Applications." *Psychological Bulletin* 147, no. 8 (August 1): 806–27.

Ober, C., Stephen Sinatra, and Martin Zucker. 2010. *Earthing: The Most Important Health Discovery Ever!,* Laguna Beach, Calif.: Basic Health Publications.

O'Mathuna, Donal. 2016. "Therapeutic Touch for Healing Acute Wounds." *Cochrane Database of Systematic Reviews,* May 3, 2016.

Oppenheimer, Gerald, and Ezra Susser. 2007. "The Context and Challenge of von Pettenkofer's Contributions to Epidemiology." *American Journal of Epidemiology* 166, no. 11: 1239–41.

Oschman, James, Gaetan Chevalier, and Richard Brown. 2015. "The Effects of Grounding (Earthing) on Inflammation, the Immune Response, Wound Healing, and Prevention and Treatment of Chronic Inflammatory and Autoimmune Diseases." *Journal of Inflammation Research* 8, 83–96.

Oster, Gerald. 1970. "Phosphenes." *Scientific American* 222, no. 2 (February): 82–87.

Pates J. 2021. "Precognitions in Elite Sports: The Role of Intuition." *Edge Science* 45 (March): 10–13.

Peifer, Corinna, and Jasmine Tan. 2021. "The Psychophysiology of Flow Experience." In *Springer EBooks,* 191–230.

Penfield W. 1958. "Some Mechanisms of Consciousness Discovered During Electrical Stimulation of the Brain." *Proceedings of the National Academy of Sciences* 44: 51.

Pennick, Nigel. 2013. *Magic in the Landscape: Earth Mysteries and Geomancy.* Rochester, Vt.: Destiny Books.

Perls, Frederick S. 1969. *Gestalt Therapy Verbatim.* Gestalt Journal Press.

Petrusich, Amanda. 2023. "Pleasant sorrows: The Mysticism of Paul Simon." *The New Yorker,* June 5, 2023, 80.

Phelan, Matthew. 2019. "'History Is Written by the Victors' Was Not Written by the Victors." *Slate Magazine,* November 27, 2019.

Picoult, Jodi. 2001. *Handle with Care.* New York: Atria Publishing.

Planck, Max. 1968. *Scientific Autobiography and Other Papers.* Philosophical Library.

Plato. 375 BCE. *The Republic,* Book VII, 514a–520a.

Plimpton G. 1985."The Curious Case of Sidd Finch," *Sports Illustrated,* April 1, 2004.

Popular Mechanics. 2004. "Finding Water with a Forked Stick May Not be a Hoax." December 6, 2004.

Powers, C. 2020. "Paul Casey on Spectator-Less Golf Events: 'I've Really Struggled with It,'" *Golf Digest*, August 7.

Pringle, Lucy, with James Lyons. 2019. *The Energies of Crop Circles: The Science and Power of a Mysterious Intelligence*. Rochester, Vt.: Destiny Books.

Rasmussen, N. 2011. "How a Generation of "Beat" Writers Burnt Out on Speed." *The Fix*, September 22, 2011.

Redfield, James. 1994. *The Celestine Prophecy: An Adventure*. Brentwood, Tenn.: Warner Books.

Reich, Wilhelm. 1951. The objective demonstration of orgone radiation, from *The Cancer Biopathy* (vol. 2 of *The Discovery of the Orgone)*, in *Selected Writings: An Introduction to Orgonomy*. New York: Noonday Press, 208.

Reich, Wilhelm. 1973. *The Cancer Biopathy*. Rangeley, Me.: Organon Press.

Reichenbach, Karl. 1854/2010. *The Od Force: Letters on a Newly Discovered Power in Nature*. Whitefish, Mt.: Kessenger Legacy Reprints.

Rein, Glen and Paolo Giacomoni. 2010. "Skin Aging: A Generalization of the Micro-Inflammatory Hypothesis," in Farage, M. A. et al. *Textbook of Aging Skin*. Berlin: Springer-Verlag, 789–96.

Rein, Glen. 1995. "The In Vitro Effect of Bioenergy on the Conformational States of Human DNA in Aqueous Solutions." *Acupuncture and Electro-Therapeutics Research* 20, no. 3–4: 172–80.

René, Lucia. 2009. *Unplugging the Patriarchy*, Crown Chakra Publishing.

Richardson, Valerie. 2021. "Publisher Blasts 'Total Media Blackout' of Robert Kennedy's Bestseller on Dr. Fauci." *Washington Times*, December 2, 2021.

Roberts, Jane. 1972. *Seth Speaks: The Eternal Validity of the Soul*. San Rafael, Calif.: Amber Allen Publishing.

Roberts, Jane. 1981. *The Individual and the Nature of Mass Events*. Englewood Cliffs, N.J.: Prentice-Hall.

Ronay, Barney. 2020. "Eerie Silence Resounds as Germany Ushers in Football's New Abnormal." *The Guardian*, May 17, 2020.

Rosa, Linda, Emily Rosa, L. Sarner, and S. Barrett. 1998. "A Close Look at Therapeutic Touch." *Journal of the American Medical Association* 279, no. 13: 1005–10.

Rosen, George. 1946. "Mesmerism and Surgery." *Journal of the History of Medicine and Allied Sciences* 1, no. 4 (January 1): 527–50.

Rosner, Gilad. 2002. "Towards an Integral Art." *Trip Magazine: The Journal of Psychedelic Culture.* Alex Grey website, July 7, 2002.

Rubik, Beverly. 2002. "The Biofield Hypothesis: Its Biophysical Basis and Role in Medicine." *Journal of Alternative and Complementary Medicine* 8, no. 6 (December 1): 703–17.

Rubik, Beverly, David J. Muehsam, Richard Hammerschlag, and Shamini Jain. 2015. "Biofield Science and Healing: History, Terminology, and Concepts." *Global Advances in Health and Medicine* 4, no. 1 suppl. (January 1).

Rubik, Beverly, and Harry Jabs. 2018. "Revisiting the Aether in Science." *Cosmos and History: The Journal of Natural and Social Philosophy* 14, no. 2 (August 26): 239–55.

Savage, Mark. 2022. "Nation Asked to Sing Sweet Caroline for the Queen." *BBC News,* May 24.

Schneider, Michael. 1994. *A Beginner's Guide to Constructing the Universe: Mathematical Archetypes of Nature, Art, and Science.* New York: HarperCollins.

Schwartz, Gary E., Linda G. Russek, and Justin Beltran. 1995. "Interpersonal Hand-Energy Registration: Evidence for Implicit Performance." *Subtle Energies,* 6, no. 2: 183–200.

Schwartz, Stephan. 2007. *Opening to the Infinite*, 3rd ed., Wash.: Nemoseen Press.

Schwarz, Jack. 1978. *Voluntary Controls: Exercises for Creative Meditation and for Activating the Potentials of the Chakras.* New York: Plume Books.

Sharaf, Myron. 1983. *Fury on Earth: A Biography of Wilhelm Reich.* New York: St. Martin's Press.

Shapiro, Arthur. 1968. "Semantics of the Placebo." *Psychiatric Quarterly* 42: 653–95.

Shaughnessy, Dan. 2020. "Hoping Former MVP Cam Newton is More Kevin Garnett than Bob McAdoo, and Other Thoughts." *Boston Globe,* August 8, 2020.

Shaugnessy, Dan. 2023. "Have Sports Celebrations Gone Too Far?" *Boston Globe* April 4, 2023. Online commenter dac44.

Shlain, Leonard. 1999. *The Alphabet Versus the Goddess: The Conflict between Word and Image.* London: Penguin.

Sierpina, Victor S., Mary Jo Kreitzer, and Eric Leskowitz. 2007. "Innovations in Integrative Healthcare Education: Spaulding Rehabilitation Hospital—the Integrative Medicine Project." *Explore: The Journal of Science and Healing.* January 1 vol. 3 issue 1: 70–71.

Silva, Freddy. 2021. "They're Alive! Megalithic Sites Are More than Just Stone." *Ancient Origins: Reconstructing the Story of Humanity's Past* website, September 11, 2021.

Sinclair, Upton. 1930. *Mental Radio: Does It Work, and How?* Self-published.

Singh, Kumari Vandana, Chandra Prakash, Jay Prakash Nirala, Ranjan Kuman Nanda, and Paulraj Rajamani. 2023. "Acute Radiofrequency Electromagnetic Radiation Exposure Impairs Neurogenesis and Causes Neuronal DNA Damage in the Young Rat Brain." *Neurotoxicology* 94: 46–58.

Smith, Esther. 2016. *Avebury Stone Circle: World Heritage Site.* UK: Forward Publications.

Smith, Rory. 2020. "Do Empty Stadiums Affect Outcomes? The Data Says Yes." *The New York Times,* July 1, 2020.

Soh, Kwang-Sup. 2009. "Bonghan Circulatory System as an Extension of Acupuncture Meridians." *Journal of Acupuncture Meridian Studies* 2, no. 2: 93–106.

Solfvin, Jerry, Daniel Benor, and Eric Leskowitz, 2005. "Questions Concerning the Work of Daniel P. Wirth." *Journal of Alternative and Complementary Medicine* 11, no. 6: 949–50.

Spall, Rafe. 2021. "I Think I Am Saying Too Much." *The Guardian,* May 2, 2021.

Stapleton, Peta, Oliver Baumann, Tom O'Keefe, and S. Bhuta. 2022. "Neural Changes After Emotional Freedom Techniques Treatment for Chronic Pain Sufferers." *Complementary Therapies in Clinical Practice* 49 (November).

Steptoe, Andrew. 1986. "Mozart, Mesmer and 'Cosi fan tutte.'" *Music and Letters* 6, no. 3 (July 1).

Stevens, Ramon. 1988. *Whatever Happened to Divine Grace.* Ojai, Calif.: Pepperwood Press.

Stevens, Ramon. 1992. *The Alexander Journal* 17, no. 1 (September).

Stevens, Ramon. 1995. "Just Say Yes: Drugs and Human Consciousness." Chap. 8 in *Spirit Wisdom: Living Consciously in an Age of Turmoil and Transformation.* Ojai, Calif.: Pepperwood Press.

Sullivan, Tara. 2020. "Dustin Johnson's Dominance a Rare Breed in a Changed Sporting World." *Boston Globe*, August 24, 2020.

Swanson, Claude. 2011. *Life Force, the Scientific Basis: Volume 2 of the Synchronized Universe*. Tuscon, Ariz.: Poseidia Press.

Swanson, Claude. 2013. "The Torsion Field and the Aura." *Subtle Energies and Energy Medicine* 19, no. 3: 23–90.

Tharawadeepimuk, Kittichai, and Yodchanan Wongsawat. 2017. "Quantitative EEG Evaluation for Performance Level Analysis of Professional Female Soccer Players." *Cognitive Neurodynamics* 11, no. 3 (February 24): 233–44.

Tiller, William A., Walter Dibble and Gregory Fandel. 2005. *Some Science Adventures with Real Magic*. Walnut Creek, Calif.: Pavior Publishing.

Tiller, William A., Rollin McCraty, and Mike Atkinson. 1996. "Cardiac Coherence: A New, Noninvasive Measure of Autonomic Nervous System Order." *Alternative Therapies in Health and Medicine* 2 (Feburary 1): 52–63.

Tompkins, Peter. 1973. *The Secret Life of Plants*. New York: Harper and Row.

Travell, Janet, and David Simons. 2018. *Myofascial Pain and Dysfunction: A Trigger Point Manual*, third ed., LWW.

Tripoli, Catt. 2016. *Conscious Fitness: Strength Training for the Evolution of Body, Mind and Spirit*. Bull Terrier Press.

Tsafrir, Judith. 2024. *Sacred Psychiatry: Bridging the Personal and Transpersonal to Transform Health and Consciousness*. Austin, Tex.: Greenleaf Book Group.

Turner, Christopher. 2006. "Mesmeromania, or The Tale of the Tub: The Therapeutic Powers of Animal Magnetism." *Cabinet Magazine* 21 (Spring).

Turner, Christopher. 2011. "Wilhelm Reich: The Man who Invented Free Love." *The Guardian*, July 8, 2011.

Upledger Institute International. 2023. Upledger website.

van der Kolk, Bessel. 2015. *The Body Keeps the Score: Brain, Mind, and Body in the Healing of Trauma*. New York: Penguin Publishing Group.

Volin, Ben. 2020. "A Deeper Look at the Top Trends for the 2020 NFL Season." *Boston Globe*, January 9, 2020.

Warburton, Greg. 2013. *Warburton's Winning System: Tapping and Other Transformational Mental Training Tools for Athletes*. Denver, Colo.: Outskirts Press.

Watkins, Alfred. 1925. *The Old Straight Track: Its Mounds, Beacons, Moats, Sites and Mark Stones*. London, UK: Methuen Press.

Weeks, J. 2020. "How the Scientologists Muzzled the AMA." *Journal of Alternative and Complementary Medicine*, October 25. John Weeks website.

Weil, A. 1972. *The Natural Mind: An Investigation of Drugs and the Higher Consciousness*. Boston: Houghton Mifflin.

Wheatley, Maria, and Busty Taylor. 2014. *Avebury: Sun, Moon and Earth Energies*. Marlborough, UK: Celestial Songs Press.

Whitley, Julie Anne, and Bonnie L. Rich. 2008. "A Double-Blind Randomized Controlled Pilot Trial Examining the Safety and Efficacy of Therapeutic Touch in Premature Infants." *Advances in Neonatal Care* 8, no. 6 (December 1): 315–33.

Wikipedia contributors. 2023. "Royal Commission on Animal Magnetism." Wikipedia, April 11, 2023.

Wilber, Ken. In Leskowitz (Ed.). 2014. *Sports, Energy and Consciousness: Awakening Human Potential through Sport*. CreateSpace.

Williams M. Wigmore T. 2020. *The Best: How Elite Athletes Are Made*. Nicholas Brearley.

Wilson E. O. 2008. *Superorganism: The Beauty, Elegance and Strangeness of Insect Societies*. New York: W. W. Norton.

Winter, A. 1998. *Mesmerized: Powers of Mind in Victorian Britain*. Chicago: University of Chicago Press.

Winter, Dylan. 2010. "Amazing Starlings Murmuration (Full HD) Keepturningleft.Co.Uk," YouTube video, November 13, 2010.

Wirth, Daniel. 1990. "The Effect of Non-Contact Therapeutic Touch on the Healing Rate of Full-Thickness Dermal Wounds." *Subtle Energies and Energy Medicine* 1, no. 1: 1–10.

Wohlleben, Peter. 2016. *The Hidden Life of Trees: What They Feel, How They Communicate—Discoveries from a Secret World*. Vancouver, Canada: Greystone Books.

Wright, Bonnie. 2019. "Acupuncture for the Treatment of Animal Pain." *Veterinary Clinics of North America Small Animal Practice* 49, no. 6 (September).

Yogananda, Paramahansa. 1946. *Autobiography of a Yogi*. Los Angeles: Self-Realization Fellowship.

Zamani, Hoda, Mohammad Nadimi-Shahraki, and Amir Gandomi. 2022. "Starling Murmuration Optimizer: A Novel Bio-Inspired Algorithm for Global and Engineering Optimization." *Computer Methods in Applied Mechanics and Engineering* 392 (March): 114616.

Index

Page numbers in *italics* refer to illustrations.